How Adult Stem Cell Therapies

SAVED MY LIFE

Medicine's Best-Kept Secret Can Save Your Life, Too

Bernard van Zyl

Rooftop Publishing™
1663 Liberty Drive, Suite 200
Bloomington, IN 47403
Phone: 1-800-839-8640

This book is a work of non-fiction. Unless otherwise noted, the author and the publisher make no explicit guarantees as to the accuracy of the information contained in this book and in some cases, names of people and places have been altered to protect their privacy.

First published by Rooftop Publishing 3/29/2007

Publisher: Kevin King
Senior Editor: Lesley Bolton
Cover Design: April Mostek
Book Design: Andrew Craig
Production Manager: Brad Collins

ISBN: 978-1-60008-037-1 (sc)

Library of Congress Control Number: 2007925068

Printed in the United States of America
Bloomington, Indiana

This book is printed on acid-free paper.

This book is dedicated to my lovely wife, Lynn, whose care and love over a five-year period saved my life at great personal sacrifice.

"Do not go gentle into that good night,
Rage, rage against the dying of the light."

Dylan Thomas

Acknowledgments

I wish to express my grateful appreciation to Dr. Douglas Spector, who saved my life during a long night on the operating table. I am also grateful to Dr. Mark Milunski and his team at the Florida Heart Group, who did everything possible to keep me alive over a five-year period. Special thanks to pioneer Dr. Douglas Losordo, who repaired my badly damaged heart and returned my life to normal. I am very grateful to Dr. K. K. Poh, Daniela Grasso, Marianne Kearney, Marisa A. Villarroel, and the many other caring and friendly people at St. Elizabeth's Medical Center in Boston.

I will forever appreciate the help of Bob and Kay van Zyl and Kate van Zyl, who made my trips to Boston possible.

CONTENTS

Appendices

Foreword

There are two great things about being a physician. The first is the ability to positively impact the lives of individual patients. Nothing compares with the thrill of seeing a patient's health improve by virtue of a treatment that has been prescribed or a procedure that has been performed. Second is the opportunity to impact the lives of many patients, both now and in the future, by developing a treatment that was previously unavailable.

There are several key ingredients that are required for the latter of these possibilities to occur, and the most essential is the courage, not of the physician investigators, but of patients who volunteer to participate in clinical trials testing new therapies. Without these true pioneers, medical research, and therefore medical treatments, would not advance. Recently, mounting evidence has documented the existence of stem cells in adults. This notion was inconceivable until recently; however, there is incontrovertible proof that our bodies harbor cells that are not only capable of repairing damaged tissue, but probably do so on a relatively routine basis. Harnessing the power of these cells has deservedly achieved major attention, as scientific evidence has mounted suggesting the potential therapeutic potency of these cells.

The road of therapeutic development, however, is long and rigorous. Early phase clinical trials must establish the safety of the new approaches before expanding to include the large number of patients required to establish the effectiveness of these therapies. All along this road, courageous and generous individuals must volunteer their time and, yes, accept the risks associated with a new and not completely understood therapy.

Anyone who had ever taken a medicine of any kind probably owes a debt of gratitude to someone like Mr. van Zyl and others in clinical trials who are willing to put themselves on the line for the good of mankind.

—Douglas Losordo, M.D., Director, Feinberg Cardiovascular Research Institute and Program in Cardiovascular Regenerative Medicine, Northwestern University Feinberg School of Medicine

Preface

Did you know that you don't have to be rich or famous to receive free advanced medical treatments in FDA-approved clinical trials? Did you also know that patients are being treated right now in clinical trials with adult stem cells that were harvested from their own bodies?

Remarkable results have recently been achieved in the treatment of almost all major diseases using adult stem cell therapy. There is little need to wait for the political outcome of the great controversy surrounding embryonic stem cells.

Why are clinical trials of adult stem cells unknown by the general public? The answer may lie in the preference of the print and broadcast media to report on the controversies surrounding embryonic stem cells to the almost total exclusion of progress made with other types of stem cells. News in the popular media concerning life-saving achievements in fighting disease with adult stem cells is very rare.

Some physicians and scientists believe that adult stem cells are not as versatile or effective as embryonic stem cells, while others have shown that adult stem cells are highly effective against most major diseases. The greatest difference at the moment is that adult stem cells are being successfully tested on human subjects while embryonic cells are still being evaluated in the laboratory and in mice. There is some evidence that embryonic stem cells cause tumors under certain conditions and that rejection may be a factor when they are implanted into human subjects. These problems have not been noted with adult stem cells.

I would not have become familiar with adult stem cells if I had not suffered a severe heart attack that turned me into an invalid waiting to die from congestive heart failure. Extensive research led me to an adult stem cell clinical trial that had the objective of rebuilding the heart using the patient's own stem cells.

The clinical trial consisted of a group of twenty-four patients with severe heart disease that could not be brought under control by conventional surgery or therapy. I was one of the twenty-four who entered into an FDA-approved clinical trial in which adult stem cells were harvested from my own body and injected directly into my failing heart. The stem cells increased blood circulation to damaged heart muscle and increased my heart's pumping capacity. I went from being a dying invalid to a normal, robust human being. I found the clinical trial by an extensive Internet search. My experience illustrates that seriously ill and dying patients can take charge of their own cases and be certain that they are receiving the best possible treatment. Doctors are limited to applying treatments that the FDA has approved for general practice after extensive testing for safety and efficacy. To receive advanced treatments that may prove to be life saving, patients must become participants in clinical trials.

My brush with death illustrates to all those who are seriously ill that they should never give up. I give instructions on how to use the Internet to find FDA-approved stem cell therapies that are in clinical trial but not yet available to practicing physicians. I offer hope to millions of patients and their loved ones that new stem cell treatments may be available to ease their suffering or save their lives. In addition, I show them how to manage their own cases so that they can find stem cell therapies for their own particular illness.

Chapter One

Stem Cells and the Heart

This book is about stem cells and stem cell therapy. Many believe that stem cell therapy may lie somewhere in the future. That is not the case, and I am living proof that stem cell therapies are already available. They are repairing hearts, rebuilding livers, and assisting in the cure of cancer and a multitude of other diseases. Popular broadcast and print media led me to believe that stem cell therapy was a long way off, held back by presidential order and political factions. I am one of the lucky ones. I found my way into a stem cell clinical trial that repaired my badly damaged heart and restored my life after I had come much too close to death for comfort. There were only twenty-four of us in the Phase I stem cell clinical trial.

We had all come for the same purpose: to ease the frightening pain in our chests and to try to stay alive. We were just a few of the tens of millions of people with similar problems who would ultimately benefit if the therapy under test proved to be successful. All twenty-four patients were very lucky to be selected for the pioneering stem cell clinical trial. In current medical practice, patients with such severe heart disease suffer chronic heart pain (angina) and die slow deaths from congestive heart failure, their lungs filling with blood that their weakened hearts could not pump out to their bodies.

We no longer face this grim prospect because of the amazing capabilities of stem cells. Uniquely, they can multiply themselves into large numbers and change themselves into any type of cells needed by the body, including brain cells, muscle tissue, and liver and kidney tissue.

As explained in later chapters, stem cells come from many sources. One source of adult stem cells is bone marrow. In our clinical trial, a newly invented biological drug coaxed stem cells from our bone marrow into our bloodstreams. From there, the stem cells were collected and later injected into our heart muscle, where they changed themselves into vascular and muscle cells. These new cells improved blood circulation to our hearts and restored heart muscle, enabling our hearts to pump enough blood to restore our health.

The Texas Panhandle seems a world away from society's great debate about stem cells. If you want to see if stem cells can transform medicine, a ranch near Dalhart, Texas is a good place to start. It is owned by Robert Young, patient number three in the pioneering stem cell clinical trial to repair damaged hearts. The way Young's life has changed indicates a breakthrough in treating heart disease, perhaps on a par with bypass surgery.

Young, at age sixty-eight, was fighting a losing battle with a particularly bad form of heart disease. His major coronary arteries quickly and repeatedly clogged after doctors had cleared them. He had undergone quadruple bypass surgery and twenty-four balloon angioplasties, a procedure in which a catheter is fed through a thigh artery to unblock arteries in the heart. His deteriorating condition forced him to give up raising cattle on his 2,300-acre ranch, and he couldn't walk more than a few steps without sharp chest pain and shortness of breath. He had run out of conventional options.[1]

Young flew to Boston to participate in the landmark stem cell clinical trial conducted by cardiologist Douglas Losordo at Caritas

St. Elizabeth's Medical Center. His own stem cells were coaxed out of his bone marrow and then injected into the damaged areas of his heart muscle. After several months, Young was a changed man, able to do almost any physical activity. He recently built a deck on his family's getaway at 8,500 feet in the Colorado Rockies.

For four years, Mildred Salas thought that every Valentine's Day might be the last with her fifty-eight-year-old husband, Anthony. After thirty-three hospitalizations and innumerable angioplasties, doctors told Salas that he had an inoperable heart condition. "Basically, doctors told me to go home and die," Salas said. Fortunately, his cardiologist, Richard Schatz at the Scripps Green Hospital in La Jolla, California, became a member of Dr. Losordo's stem cell clinical trial team, and Salas was able to become number six in the study.

What has happened to Salas, according to Dr. Schatz, "…is amazing. He feels like he has a new life now."

"I feel great; I don't have any pain, and I don't have to use oxygen, nitroglycerin, and morphine anymore," says Salas.[2]

I was number nineteen in the study, and my story is typical of what can happen when the heart goes bad. In the past, I, too, would have died with conventional care available to physicians before the advent of stem cell therapy.

The troubling signs had been there, but my wife, Lynn, and I were too busy with the upcoming holidays to take them as seriously as we should have. For a few weeks, I had been complaining of congestion in my chest, but I didn't seem to have a cold. I was sure that the congestion was not a sign of anything serious since I had had a complete physical at the Mayo Clinic in 1981 and received a good report on my heart. However, since it was now 2000, and I was sixty-seven years old, Lynn wanted me to get an updated heart work-up.

I made an appointment with cardiologist Mark Milunski, MD, of the Florida Heart Group, who examined me and scheduled a stress test for December 18. It would be too little, too late.

On Sunday, December 17, we were up at about nine as we had a lot to do this last full weekend before Christmas. I was still congested, so I went to see a doctor at a walk-in clinic for what I thought must be a cold or perhaps walking pneumonia. The doctor suggested several over-the-counter cold remedies, which I bought on the way home. If only the doctor had dug deeper into the cause of my shortness of breath, things might have turned out differently.

When I got home, I took the cold medicine and went to lie down. Suddenly, I felt a terrible pain in the center of my chest. It was as if someone had hit me with an axe. I walked into the kitchen and told Lynn that I needed to go to the hospital. She said she would change clothes and drive me, but I replied emphatically, "No, I must go right now!" We got into the car and drove to the hospital, which was only a half a block away at the end of our street.

Lynn dropped me off at the door to the emergency room and went to park the car. When I went inside, a triage nurse saw me and immediately took me into the emergency room, helped me onto a gurney, and started to take an EKG of my heart. In the meantime, Lynn parked the car and was in the lobby filling out insurance forms. A nurse came over and asked, "Are you Mrs. van Zyl?" When she said she was, the nurse calmly told her, "Your husband's heart stopped, but we were able to restart it. I'll take you back to him right away."

Lynn was speechless for a moment and then asked incredulously, "You mean he's having a heart attack?" She couldn't believe it!

I had suffered a cardiac arrest and died. The emergency room doctors and nurses brought me back to life by shocking my chest with those electric paddles you see on TV. The doctors said it was a miracle that I survived; help is usually not immediately available, and the patient is dead within minutes. Doctors call this type of sudden cardiac arrest "the Widow maker." Even in the emergency room, the chance of reviving a patient is only 10%.[3]

I remember a nurse doing the EKG, but then I suddenly fell asleep and immediately started dreaming. I never remember my dreams, but I know that this one was very pleasant. Suddenly, I was aware of people shouting, "Bernie, wake up! Wake up!"

A doctor looking closely into my eyes said, "Your heart stopped, but we managed to get it started again. How do you feel?" I was very lucid, and I told him I was okay.

Once in the patients' area, Lynn saw several nurses and emergency doctors surrounding someone. As she came near, it became obvious this was very serious. I was covered with sweat and suffering intense pain. A doctor related that they had just given me an EKG, which looked normal, when my heart stopped. He had shocked my heart back to beating, but I was still in a great deal of pain. When I saw Lynn next to me, I told her that I had died, but I didn't see a bright white light to follow.

Lynn gave them Dr. Milunski's name at the Florida Heart Group and told them that I was scheduled for a stress test the next day. One of the nurses on duty, who worked for the Florida Heart Group during the week, was able to get in touch immediately with the on-call doctor. He said he was on the way to another hospital but would detour to Winter Park to see me. Luck was with us. Dr. Kerry Schwartz, a specialist in electrical problems of the heart, immediately sent me by ambulance to Florida Hospital South in Orlando where, I was to learn later, they do more coronary bypass operations than any other hospital in the country.

When Lynn arrived at the catheterization lab, the nurses told her that I was already having a heart catheterization which Dr. Schwartz had called ahead to set up. She was sent to the waiting room where she called all of the members of our family that she could reach. My daughter, Kris, was there in less than a half hour. Lynn said later that when Kris walked into the waiting room, she was so relieved to have company. After a few minutes, the doctor came, took them to a diagram of the heart, and showed them what had happened to me. He said the left main artery to my heart was completely blocked. I needed an immediate bypass operation to provide blood to the part of my heart muscle that was now starving for oxygen. Lynn and Kris were brought into the cath lab where I was lying on the table looking very pale. The doctor introduced them to the surgeon, Dr. Douglas

Spector, whose clothes reflected that he had just been called in from cutting his lawn. He showed them the catheterization video on a small screen. The blockage was all too clear in the eerie blue of the film. Dr. Spector explained that I needed surgery immediately because my heart had stopped beating again during the catheterization procedure and had to be shocked back into beating.

Because of the pain medication, I didn't fully realize what was happening and demanded to go home, as I had appointments the next day. Dr. Spector gently told me, "Mr. van Zyl, if you try to walk out of the hospital, you will be dead before you reach the door." Lynn gave her consent for the bypass operation and persuaded me to be taken up to surgery. She and Kris went to the surgical waiting room. This room became the center of their universe for the next twelve hours.

Lynn called her daughter, Cathy, in Virginia again to let her know that I was in surgery. Cathy, a pharmacist, told Lynn that bypass operations had become everyday events, which made Lynn more optimistic about the outcome. She stayed calm and was able to reassure and comfort other family members as they started to arrive.

During the past year, Lynn had been a casual observer of the wonderful results of bypass operations. At her previous job, the manager and his father each had had bypass operations and were back to work in six weeks. Only two months before, my nephew had had a heart attack which required bypass surgery. After surgery, most bypass patients were in cardiac intensive care for three days before going to a progressive care unit for several more days. They then returned home and went to cardiac rehabilitation for six weeks.

Other family members arrived, and Lynn told them she expected they would all be home before midnight. She was wrong; this was to be the longest night of her life.

At about 3 a.m., a weary Dr. Spector came into the room and spoke to them. He told them that my heart had stopped again before I was put on a heart-lung machine for the bypass operation. A heart-lung machine provides oxygen exchange in the blood and pumps blood through the body during the operation while the heart is motionless. Dr. Spector said that he did six bypass grafts, but when he had finished, my heart would not start beating on its own. He tried three times without success. Dr. Spector said he would leave me on the heart-lung machine a little longer to rest my heart and then try one more time. Lynn sensed at that moment that he was trying to prepare her to lose me.

The family knew that the situation was very bad. Lynn talked to Cathy again who assured her that surgeons don't like to lose patients on the operating table and that the doctor would do his very best to get me to recovery. Lynn could see that the family members were losing hope, and then she realized that she was still holding a little bag with my watch and wedding ring that a nurse had given her earlier. She put them on and had the family hold hands in a circle and project their joint strength and love to me while visualizing my heart beating and repeating aloud, "Breathe, Bernie. Breathe, Bernie..."

Sometime before 5 a.m., Dr. Spector came out to tell them that my heart had suddenly started to beat! He said that they would be able to see me for a few minutes after the nurses got me to the cardiac intensive care unit. He explained to them that I would be hooked up to many machines including a heart pump to assist my heart until it could beat strongly enough on its own. Physically and emotionally exhausted, they hugged each other and nodded silent thanks. They were told that the next forty-eight hours were critical.

When Lynn and Kris came to the cardiac intensive care unit, they were told that they could stay only five minutes. Nothing could have prepared them for what they were to see.

There were tubes and wires entering, exiting, or attached to the unrecognizable person on the bed. I had become a huge, bloated carcass with a face double its normal size. My arms, legs, and torso were swollen beyond recognition. My inner thighs had stitches from the groin to just above the knees and below the knees to my ankles. With the thumping sound of a large bellows, the big machine at the foot of the bed assisted my heart by circulating my blood through a tube running into a groin incision. Wires were attached to my chest, which had been split open during surgery. A ventilator tube in my throat kept me from talking, and I barely recognized that they were there. Monitor screens overhead showed a myriad of lines and numbers. A partially full urinal bag hung from one side of the bed, and multiple IV bags were clustered on hooks above the bed. Clear bins of assorted needles, medicines, and gauze pads were on the shelves and counters. A window revealed a patch of sky just starting to lighten. This could not be real! Had I fallen into some horrible *Twilight Zone* episode?

They told me they loved me and assured me that I would be okay. Gingerly, they kissed me before they had to leave. Lynn was given a slip of paper with the phone number and visiting hours: 9:00 to 9:15, 11:00 to 11:15, 1:00 to 1:15, 3:00 to 3:15, and 7:00 to 7:15. These times would govern her universe for the next days.

When Lynn got home, she showered and lay down for a little while, but sleep would not come as she wanted to be sure to leave early enough for the drive back to the hospital for the 9:00 a.m. visit. She left at 8:30 with a piece of toast and the first of many chewable Pepsids that would be her diet for the next few days. Once she reached the hospital waiting room, visiting time couldn't come soon enough. When she came in to see me this time, the shock of seeing the sterile equipment attached to me was gone, and she could touch my arm at one of the few spots without an IV, bandage, or wire. While she was there, a nephrologist came in to check my kidney function. He joked to the nurse about the surgeons putting patients into kidney shutdown, which kept him in business getting them going again. Then he added some medication and was off to the next patient. Later, I was to learn that during a heart attack, the body saves the heart and the brain by shutting down blood flow to the extremities and all other organs. This lack of blood to the organs and the rest of the body can cause damage that doesn't always show up immediately, as we were to find out.

I was awake all day but still groggy. Seeing me awake allowed Lynn to be calm again, taking the kidney news in stride and still expecting all to be well. The new nurse was very expert at adjusting medications and had a phone to put her in instant touch with Dr. Spector or his assistant surgeon. After operating all night, Dr. Spector still had a full day of surgery but found time to stop in often to check on me. My cardiologist, Dr. Milunski, came in and assured Lynn that my condition was normal after the bypass surgery. Neither Lynn nor I understood that the doctors didn't expect me to live.

When Lynn and Kris came in for the 7:00 p.m. visit, I was wide awake and very interested in everything that was going on around me. Lynn was amazed that I could be so alert after everything I had been through in the last thirty hours. With the respirator tube still in my throat and both swollen arms full of IVs, I could hardly move to write questions, so I tried to spell out words in Lynn's hand with my finger. With her lack of sleep, I frustrated us both as I tried to spell faster and she understood less. Kris traded places with her and was able to decipher what I wanted. As Kris repeated each letter "w-h-o-w-o-n," Lynn was still confused.

Finally, when Kris answered, "Bush," it dawned on Lynn that I wanted to know if the Supreme Court had announced a decision on the presidential vote in Florida. They both laughed that I even remembered that a decision was due today. My mind seemed to be intact.

When Lynn went home after the 7:00 p.m. visit, the voice mail on the phone was full of calls from family and friends wanting updates. There was also a message reminding me of my stress test! After making other calls, Lynn spoke with Cathy. Cathy's clinical questions were beyond Lynn's understanding at this point, but for the first time, Lynn told her about the fear in her stomach that drove her retching to the bathroom at home every few minutes. Her morning toast was still sitting in her car. She could only chew Pepsid, as she seemed unable to swallow food. She had never before revealed any weakness to her children, and soon after that call, Cathy called back to let her know that she would be down on the next available flight.

Lynn was finally able to get to sleep by wrapping up in my bathrobe and lying on my pillow. My scent, lingering on my robe and pillow, comforted her. Suddenly, the phone was ringing. Struggling to wake up in the darkness, she grabbed the phone, terrified. My night nurse was on the line asking if she could come in to help settle me down. The nurse indicated if she could not, he would have to restrain me. She thought, "How could this be?" When she had last seen me, I was alert and calm. Bewildered, she drove back to the hospital. The streets were nearly deserted and the hospital quiet as she made her way to the CICU. As she entered my room, she could see I was awake and agitated. The nurse told her that some medications can cause confusion.

I waved her over to me and motioned for a pad. "Take me home," I wrote. "He is trying to kill me!" Even with all the pain medication I had been given, I had not slept for several days, and I insisted to the nurse that I did not want to sleep. I couldn't help thinking of the cardiac arrest when I seemed to have fallen asleep and was dreaming, while I was actually dying. I didn't want to fall asleep again because it might be another cardiac arrest, so I fought sleep with every fiber of my mind, body, and soul. I don't know how many days I managed to stay awake, but the nurse had become concerned and was trying to medicate me so that I would sleep. I thought he was trying to kill me! I told him so and became very angry, fighting him off as best I could. He must have called and asked Lynn to come in, because she was there very quickly.

The nurse pulled a recliner next to my bed, and Lynn lay next to me, put her arm across my chest, and whispered quietly to calm me and get me to try to sleep. Lynn has a wonderful way of soothing and comforting me. She held me and whispered that it would be okay if I slept, that she would not let anything happen to me, and I dozed off uneasily. But, by morning, Lynn was totally exhausted.

For many months, I remained fearful of falling asleep.

When the day nurse came on, I seemed able to sleep, and Lynn was sent out with the news that I would not be allowed visitors due to my confusion and agitation the night before. Lynn came out to find my sister and her husband in the waiting room. Sadly, she had to tell them about the restriction on visitors.

Once at home for the evening, Lynn had the same rounds of calls to return with the day's updates. She learned that my brother, Bob, and his wife, Kay, would be flying in from Philadelphia on Wednesday afternoon and would go straight to the hospital to meet her there. Cathy would arrive at the airport Wednesday morning. Cathy's expertise was a blessing. As soon as she arrived, she put her cardiology training to use, asking the right questions to obtain accurate information. She very thoughtfully asked how the doctors felt about one medication over another and made sure of the reasons for the use of each. Later, Lynn relaxed a little as Cathy explained the situation. Lynn wanted Cathy to be completely candid as to what she could expect and what my chances were. While Cathy said that the medications I was taking yielded the best two-year survival rate, she was not encouraging.

I was weaned from the respirator that day, but my heart was still not strong enough to operate without the heart pump which continued to pump along for many more days.

My brother and Kay were able to get to the hospital in time for the 7:00 p.m. visit on day three. By then, much of the swelling was down thanks to the restoration of my kidney function, and without the respirator tube, I thought I looked pretty good. However, Bob was shocked to see all the tubes and how swollen I still was. After the visit, Bob and Kay graciously took Lynn and Cathy out to dinner to give Lynn a break. The next day, Bob and Kay came to the 11:00 a.m. visit. While they were in the room, the surgeon came in to check on me and talked to Bob about what had happened and about having to shock my heart a couple more times to re-establish a proper rhythm. As they talked, Dr. Spector suggested a doctor in Philadelphia that Bob should see due to our family history of heart problems. This was to make a big difference to Bob, as his cardiologist at home discovered that Bob also had an arrhythmia problem. Fortunately, he responded to the simple treatment needed.

After seven days, my heart needed less help from the heart pump, but I could not leave the CICU until it was no longer needed. Most patients need to spend only three days in CICU.

After ten days, on December 27, I was taken off the heart pump. That day, when Lynn was leaving after her evening visit, she bent over to kiss me goodbye and noticed a terrible smell. My nurse said I would be getting a shower for the first time the next day and would smell better after that. The next day, I was moved to the progressive care unit where I could get out of bed to begin cardiac rehabilitation exercise by walking around the hallway with my heart pillow clutched to my chest. Heart pillows sound funny. They are colorful pillows made in the shape of hearts by volunteers. Open-heart-surgery patients clutch them to their chests to ease the pain caused by the breastbone, which has been sawn in half and is now held together by twisted wires.

Lynn and I both thought that I would now be home in a few days.

In the PCU, I shared a room with a younger man who had come to the hospital with chest pain. I missed the quiet of the CICU, as my roommate had a half dozen children and other family members in the room at all times. However, the noise and confusion gave me a good reason to go out for walks around the floor. Alive, but weak, I was hoping to be home by New Year's Eve. On Friday, December 30, Lynn was expecting that I would call to tell her that I would be released that day or the next. However, that was not the phone call she received. I had to tell her that I had suddenly spiked a fever of over 101°. We both knew that hospitals don't send post-operative patients home with a fever, because it could be an indication of infection.

When Lynn got to the hospital, my roommate had been moved to another room, and I was in isolation. The fever was raging, and I must not have looked good at all. The doctors decided to send me for an MRI body scan to see where the infection was hiding. Although I had always been an adventurer and had flown with bush pilots in Alaska and climbed Mounts Kenya and Kilimanjaro in Africa, I resisted going for the MRI because I couldn't stand the thought of being confined inside the narrow tube of an MRI machine.

To this day, I cannot explain how or why I developed claustrophobia during the heart attack. I had had no fear of being in an MRI machine in the past when I had a problem with arthritis.

However, the thought of going into that small tunnel in the machine now terrified me! Again, Lynn, with her ability to soothe me by just her touch, saved the day.

Only by promising to hold my hand the whole time was she able to convince me that the MRI could be done. How funny it must have looked to see me inside the small MRI opening, while Lynn stood on the outside stretching her arm inside to keep her hand on my shoulder while the MRI clunked away.

The news was bad. The insides of both thighs had become infected with staphylococcus bacteria where the veins had been harvested and used as grafts in my heart to pass blood around the blocked arteries. Large pools of infection floated inside my legs, and I needed massive doses of antibiotics through my IVs. However, the antibiotics were damaging my veins, which began to collapse. Because my veins were no longer usable, a special team put in a PIC line, which was a special IV line directly into my heart. An infectious diseases specialist was also called in, and the year 2000 passed into 2001 with me still in the hospital.

The infection in my legs continued to worsen. The doctors made a decision to operate and open my legs to remove dead flesh and drain the growing pools of infection. I would have to be anesthetized in the operating room for the procedure. This was a frightening experience for us and for the surgeon, as I was a very poor surgical risk. After they operated, they left my leg wounds open so that the infection could drain. When I returned to my room after surgery, the nurses showed us that my legs were packed with absorbent gauze that needed to be removed and the legs irrigated with antiseptic three times a day. Due to the infection, I was no longer permitted to walk to the shower or exercise in the halls.

Changing of the packing became an ordeal that I didn't complain about, but the nurses had to give me IV morphine before each procedure. Two nurses were required each time; one to unpack, clean, and repack, and the other to hold the wounds open and pass sterile instruments and gauze. Altogether, it took nearly a half hour to accomplish the procedure. I wish I could tell you that it was over in a week, but this went on for another three weeks. I became quieter as the days passed, refusing all visitors except my family. I didn't want even them to know what I was going through. Altogether, my legs were operated on four times.

I tried not to show it, but I was terrified that I would lose my legs to the infection. The operations themselves were not too bad, but each time they operated, more flesh was lost. When I looked down at my legs when the nurses were repacking them with gauze, I could see almost down to my bone through openings that had spread open about four inches wide. There were times when I didn't think I would have much left after the doctors were through.

Lynn stayed with me all day on the weekends and each evening until visiting hours were over. She felt helpless and began to walk me to the sink in my room every day to give me a sponge bath and check me over. Eventually, after watching the repacking procedure many times while holding my hand, Lynn was able to glove up and help to hand the gauze and hold my skin flaps open if a second nurse was not available. The first time she saw the yellow globs of fat and the flesh cut nearly to the bone, she thought she would pass out, but she is a real trooper and was determined to do anything required to help me get better. Eventually, the staph infection started to come under control, but the antibiotics were killing good bacteria as well as bad. This caused a fungus infection that covered the inside of my legs. We were told that the fungus could easily be controlled over a few days but that the medication might cause liver damage. With my leg wounds open, I was constantly oozing blood, and my blood levels and blood pressure fell alarmingly, so every other day, I received two units of blood. Ultimately, I had to go into a hyperbaric chamber that helped the healing by forcing pressurized oxygen into the wounds. Finally, at the end of January 2001, I was going home! Everything would go smoothly now, we thought. Life would get back to normal in a few weeks. Within a few short months, we discovered that this was unwarranted optimism.

Chapter Two

Continuing Problems and Treatment

After only four months, I was suddenly hospitalized again with severe chest pain. How could this happen so quickly? After some tests, Dr. Milunski explained that with six bypass grafts, my heart had almost all new plumbing but that plaque dislodged during heart surgery might be clogging minor coronary arteries and causing pain. This plaque could not be removed because the arteries were too small to be accessed. Due to my fragile condition, Dr. Milunski had been waiting before prescribing a cardiac rehabilitation exercise program which might have helped to restore circulation.

Another well-known cardiologist, who teaches at the University of Florida Medical Center, was brought in to consult. He thought I should go ahead and start exercising in a cardiac rehab program in the hope that exercise would help to establish new blood vessels around the clogged ones. This, called collateral circulation, could reduce the pain by providing additional oxygen to starved heart muscle.

The cardiac rehab program at Florida Hospital involves mild exercises under the strict supervision of a nurse. There are a series of sessions three times a week for eight weeks. Blood pressure is measured before the beginning of the exercises, at several times during the session, and at the conclusion. Stretching exercises are performed first, followed by treadmill, stationary bicycling, mild weight lifting, a repeat of the stretching exercises, and a cool-down period.

From the very beginning, aerobic exercises like stationary bicycling were a problem. The result of aerobic exercise should be increased blood pressure, as the heart needs to pump more blood to keep up with higher demands of the muscles. Unfortunately, my blood pressure dropped during the aerobic exercises. This was especially a problem because my blood pressure had been quite low following the bypass operation and had not returned to normal. The nurses at the hospital were very knowledgeable and supportive, but they halted my workouts because of the drop in my blood pressure. Therefore, I was unable to continue the cardiac rehab program, and collateral circulation to my small clogged arteries was therefore not established.

I discussed other alternatives with Dr. Milunski, who told me about transmyocardial laser revascularization, also called TMLR or TMR.[4] Like every other organ or tissue in the body, the heart muscle needs oxygen-rich blood to survive. The heart gets this blood from the coronary arteries. When the arteries are clogged or diseased, not enough blood is delivered to the heart. This is called ischemia. Cardiac pain (angina) is a signal that not enough blood is reaching the heart. My pain was sharp in nature, but other patients feel a squeezing or suffocating sensation in their chests. The pain usually occurs when the heart has an extra demand for blood, such as during exercise, after eating, or at times of stress. This is called stable angina. TMLR cannot cure coronary artery disease, but it may reduce the pain, or angina.

TMLR is a type of surgery that uses a laser to make tiny channels through the heart muscle in the lower left chamber of the heart (the left ventricle). The left ventricle is the heart's main pumping chamber.

After TMLR, when oxygen-rich blood enters the left ventricle, some of that blood can flow through the tiny channels and carry much-needed oxygen to the starving heart muscle.

No one really knows why TMLR helps reduce the pain of angina. Some doctors think that TMLR helps the growth of tiny new blood vessels in the heart muscle wall. This process is called angiogenesis. These new blood vessels bring more blood to the heart muscle, making it healthier. Others think that the TMLR laser destroys some of the pain-causing nerves in the heart muscle. Still others think that patients feel a placebo effect. This means that patients feel better because they received treatment, not because the treatment really worked.

Doctors came up with the idea for TMLR by studying the hearts of alligators and snakes, where blood to feed the heart muscle goes straight from the ventricles and into the muscles, not through the coronary arteries. Doctors thought that this might work in humans, too.

TMLR is a surgery, but it can be done while the heart is beating and full of blood. That means that a heart-lung machine is not needed. In addition, surgeons do not cut open the chambers of the heart, so TMLR is not open-heart surgery.

The operation will usually be scheduled at a time that is convenient for the patient and the surgeon, except in emergency cases. As the date of the surgery gets closer, patients are told to tell their surgeon and cardiologist about any changes in their health. If they have a cold or the flu, it can lead to infections that may affect recovery. Patients should report fever, chills, coughing, or runny nose. Most patients are admitted to the hospital the day before surgery or, in some cases, on the morning of surgery.

An anesthetic makes the patient sleep during the operation. Tests before surgery include an electrocardiogram (EKG), blood tests, urine tests, and a chest x-ray to give surgeons the latest information on the patient's health. A mild tranquilizer may be given before the patient is taken into the operating room.

Small metal disks called electrodes are attached to the chest. These electrodes are connected to an electrocardiogram machine, which monitors heart rhythm and electrical activity. An IV line will be inserted in a vein. The IV line will be used to give anesthesia before and during the operation. After the patient is asleep, a tube will be inserted down the windpipe and connected to a machine called a respirator, which will take over breathing. Another tube will be inserted through the nose and down the throat into the stomach. This tube will stop liquid and air from collecting in the stomach, to avoid sickness and bloating when the patient awakens. A thin tube called a catheter will be inserted into the bladder to collect any urine produced during the operation.

First, the surgeon makes a cut in the left side of the chest to get to the heart's left ventricle. Then the surgeon uses a special carbon dioxide laser to make thirty to forty channels in the heart muscle. These channels are about one millimeter wide, or about the size of the head of a pin. The surgeon makes the channels while the heart is pumping blood because that is when the heart's walls are the thickest and the least likely to be damaged.

The channels will bleed for a few seconds, but the bleeding will stop when the surgeon presses lightly on the channels with a finger. The tops of the channels close over with a blood clot, but inside the heart, the channels stay open. The TMLR procedure usually takes about two hours.

Sometimes, one part of the heart can be treated with bypass surgery while another part of the heart can be treated with TMLR. In these cases, TMLR and bypass surgery are done at the same time. Hospital stays are between four and seven days after TMLR. Recovery takes a long time, requiring rest and limits on activities. Doctors may then prescribe an exercise program or a cardiac rehabilitation program.

Most of the time, the symptoms of coronary artery disease and angina get better after TMLR, but it may take three months or more to see improvement. After recovery, activities may be resumed that once caused pain. Pain medications, including nitroglycerin, may be reduced.

Because TMLR is a new type of surgery, no one knows exactly what its long-term effects may be. Studies have shown that one year after surgery, 80% to 90% of patients treated with TMLR feel better than they did before the surgery. They also have a lower risk of heart attack.

Another type of laser revascularization is percutaneous transmyocardial revascularization (PTMR). PTMR is performed by a cardiologist in the cardiac catheterization laboratory. The area around an artery in the leg (the femoral artery) is numbed with anesthesia. The cardiologist inserts a long, thin tube called a catheter into the artery. The catheter is threaded through the artery and up to the heart. The doctor feeds the laser through the catheter and uses it to create tiny channels in the heart muscle. Because PTMR only requires a tiny incision at the site of the artery, both the surgery time and recovery time are shorter.

Not all patients can have the percutaneous form of laser revascularization. Both types of procedures, TMLR and PTMR, are a last resort for treating angina.

Unfortunately, my cardiologist felt that my heart was in too poor a condition for either type of laser procedure, and other treatments had to be explored.

In July, I had another heart attack because of new blockage in one of the major arteries that had previously been repaired. This time, I was treated with angioplasty, in which a catheter was inserted into an artery in my leg and threaded through to arteries to my heart. A balloon on the end of the catheter opened to enlarge the diameter of the artery and a metal stent, or mesh, was inserted to keep the artery open.

It was still not possible to get into and open the smaller arteries, and I began to require morphine around the clock for the constant pain. Dr. Milunski now thought that it was time to explore a heart transplant and made a referral to the Shands Medical Center at the University of Florida in Gainesville.

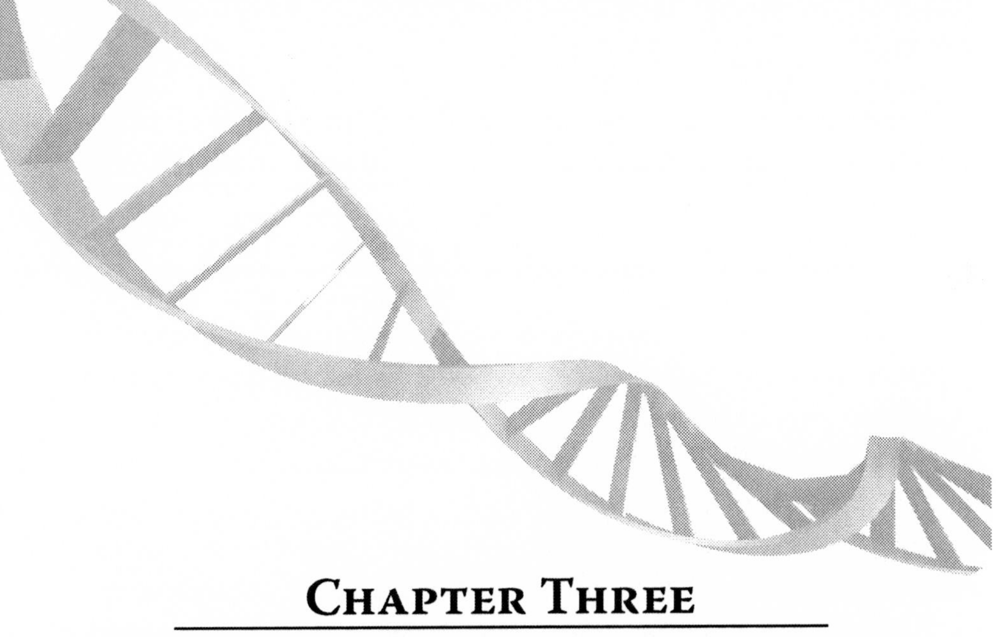

Chapter Three

Search for a Heart Transplant

The extreme gravity of my condition was brought home to me when Dr. Milunski referred me to Shands Medical Center at the University of Florida in Gainesville, Florida as a possible heart transplant recipient. I was grateful for the referral; usually, a patient cannot enter into a transplant program without one. I was about to embark on a new and frustrating journey into the world of heart transplantation.

Before an appointment could be set for a first visit, a finance department representative called to ask how I would pay for an initial physical evaluation and for a subsequent heart transplant if I were found to be acceptable. Without adequate health insurance, or personal funds for the initial work-up and possible heart transplant, it is not possible to get an appointment.

It is easy to understand the hospital's concern. The average cost of a heart transplant in the United States in the year 2000 was $148,000. (My stay at Shands for just the evaluation cost nearly $25,000.) This was not a problem, since I had group insurance as a primary payer and Medicare as a secondary payer.

Lynn and I learned that once a person is accepted as a transplant candidate, he or she is registered on UNOS (United Network for Organ Sharing) with a status code listing. This listing indicates the health of the patient or medical urgency for the needed organ. The most urgent of the heart transplant candidates have hearts that are not able to pump on their own and are on mechanical heart pumps in the hospital. A patient can only receive a heart pump if he is a candidate for a new heart, as the pump is approved for use while waiting for a transplant. Other criteria for listing were blood type, genetic makeup, size, and age. Each center had its own criteria for listing patients.

After the initial visit, the heart surgeon scheduled an angiogram for the last week of September 2001. Once into the angiogram, the surgeon decided that I had yet another blockage of an artery and balloon angioplasty and a stent were needed immediately. However, because of my heart's fragile condition, he felt such a surgery would be too risky without a backup plan in case my heart was damaged during the procedure.

The only backup could be a transplanted heart. During an emergency meeting, it was decided that I would be placed on the transplant list for the day of the angioplasty only.

I was admitted into the hospital for a complete evaluation. All of my organs were checked to detect any infection or cancer. There could not be cancer anywhere in the body as immuno-suppressant drugs would allow the cancer to flourish. Lynn and I were given psychological interviews to ensure that we were mentally ready for a transplant and for the intense after-care that would be required. There were also interviews with our back-up caregivers to determine that if I received a transplant, I would be cared for twenty-four hours a day after being released from the hospital. An accepted candidate must live within a few minutes of the hospital for months after the transplant. We were able to provide multiple backups within the family.

Fortunately, it was determined that all of my other organs were functioning well with the exception of my thyroid gland, which had been affected by my heart medications.

The angioplasty proceeded without heart damage, and the doctor was able to put in two stents to keep my arteries open. Nevertheless, some of the smaller arteries remained clogged. As a result, I still had chest pain. We had very mixed emotions about the outcome. What if my heart was damaged during the operation? If it had been, I would have been put near the top of the transplant list.

We got very bad news before leaving Shands. The transplant team decided that I would not be placed on the transplant list, because I had passed their cutoff age of sixty-five years, and they showed concern that I would have healing complications as had happened before when a staph infection had gotten into my legs. Why in the world didn't they tell me of their age limitations before I underwent a very expensive and time-consuming evaluation?

We went home desperate to try anything, as the doctor at Shands thought I might only have a few months before total heart failure.

To ease my pain, Dr. Milunski suggested another procedure called EECP (for Enhanced External Counterpulsation). This procedure offers a possibility of creating collateral circulation of the blood around small blockages.

EECP is a mechanical procedure in which long inflatable cuffs (like blood pressure cuffs) are wrapped around both of the patient's legs. While the patient lies on a bed, the leg cuffs are inflated and deflated with each heartbeat. This is accomplished by means of a computer, which operates off of the patient's EKG, so the cuffs deflate just as each heartbeat begins and inflate just as each heartbeat ends. When the cuffs inflate, they do so in a sequential fashion, so that the blood in the legs is pushed upward toward the heart.[5]

EECP has two potentially beneficial actions for the heart. First, the pushing action of the cuffs causes blood to flow toward the heart and increases blood flow to the coronary arteries. (The coronary arteries, unlike other arteries in the body, receive their blood flow after each heartbeat instead of during each heartbeat.) EECP effectively "pumps" blood into the coronary arteries. Second, by its deflating action just as the heart begins to beat, EECP creates something like a sudden vacuum in the arteries, which reduces the work of the heart muscle in pumping blood into the arteries. Both of these actions have been known to reduce cardiac ischemia (the lack of oxygen to the heart muscle) in patients with coronary artery disease.

While a primitive form of external counterpulsation has been around for a long time, it has not been very effective until recently. Thanks to new computer technology that allows perfect timing of the inflation and deflation of the pressure cuffs and produces a pushing action of blood toward the heart, modern EECP has been greatly enhanced.

EECP is administered in a series of outpatient treatments. Patients receive five one-hour sessions per week, for seven weeks. The thirty-five one-hour sessions are aimed at provoking long-lasting beneficial changes in the circulatory system. EECP now appears to be quite effective in treating chronic stable angina. (Stable angina occurs during exercise or stress and is usually relieved by nitrates.) A randomized trial with EECP, published in the *Journal of the American College of Cardiology*, 1999, showed that EECP significantly improves both the symptoms of angina (a subjective measurement) and exercise tolerance (a more objective measurement) in patients with coronary artery disease. EECP also significantly improved "quality of life" as compared with placebo therapy. More recent data show that improvement of symptoms following a course of EECP seems to persist for up to five years. Furthermore, there is preliminary data suggesting that EECP may be useful for treating unstable angina (unstable angina occurs at rest or is unrelieved by nitrates) as adjunctive therapy after angioplasty, stent, and/or bypass therapy.

The first time I underwent treatment with EECP, my chest pain diminished to a significant extent. However, after a period of about one year, the pain began again. I took another series of treatments with less effective results than the first treatment. After another year, I tried EECP again. This time, it seemed to have little or no effect.

After my first EECP treatments, I got another potential opportunity for a heart transplant. As it does almost annually, Lynn's insurance plan at her office changed to a new insurance carrier. The new insurance company preferred the heart transplant center at Jackson Memorial Hospital in Miami, Florida. We were delighted to find that Jackson Memorial has an age criterion of seventy years! Dr. Milunski immediately referred us to Jackson Memorial, and we went for an initial visit in June 2002. This time, the heart surgeon examined me and decided that he wanted to do an angiogram to determine the condition of my heart before doing any other tests.

We scheduled the angiogram for the next week and went back to Miami, expecting a three-day visit for the angiogram and work-up, but fate intervened. The day of the angiogram, I had a fever and a swollen leg. A week before, I had fallen in the bathroom and bumped my head and leg. I immediately had an MRI which showed that there was no bleeding in my brain, but, because of my poor circulation, the bruised area on my leg had become infected. The angiogram had to be postponed, and I was admitted into an isolation room on the transplant floor and given IV antibiotics.

Lynn returned home to work for a week while my infection cleared up and I became safe for the angiogram. When it was performed, the doctor found another clogged main artery and put in yet another stent to keep it open. Unbelievably, he felt I was too well to be put on Jackson Memorial's transplant list. He said he would like to see me again in five years! Perhaps the EECP had done its job so well that I had temporarily improved.

Although a heart transplant was not available for me, information concerning heart transplant may be of value to those with serious coronary disease. This information is located in Appendix B.

Looking back, I'm very happy that I couldn't get a heart transplant as I would have been tied to expensive immuno-suppressive drugs for the rest of my life to prevent rejection of the new heart. The adult stem cell therapy, which I later received, repaired my heart, and I am not threatened with rejection because the stem cells are my own.

After leaving Jackson Memorial Hospital, all was good for about six months. I was off of all pain medications. Then, the severe pain started again because my heart arteries had once again clogged. While I was in the hospital in Orlando waiting for yet another angioplasty, Dr. Milunski discovered that I had an electrical irregularity and must have a pacemaker and defibrillator implanted into my chest to regulate my heartbeat and shock my heart into life in the event that it suddenly stopped. I was going downhill rapidly.

It was at this point that I decided to search for an advanced treatment in a clinical trial since there were no conventional treatments that could be of use to me.

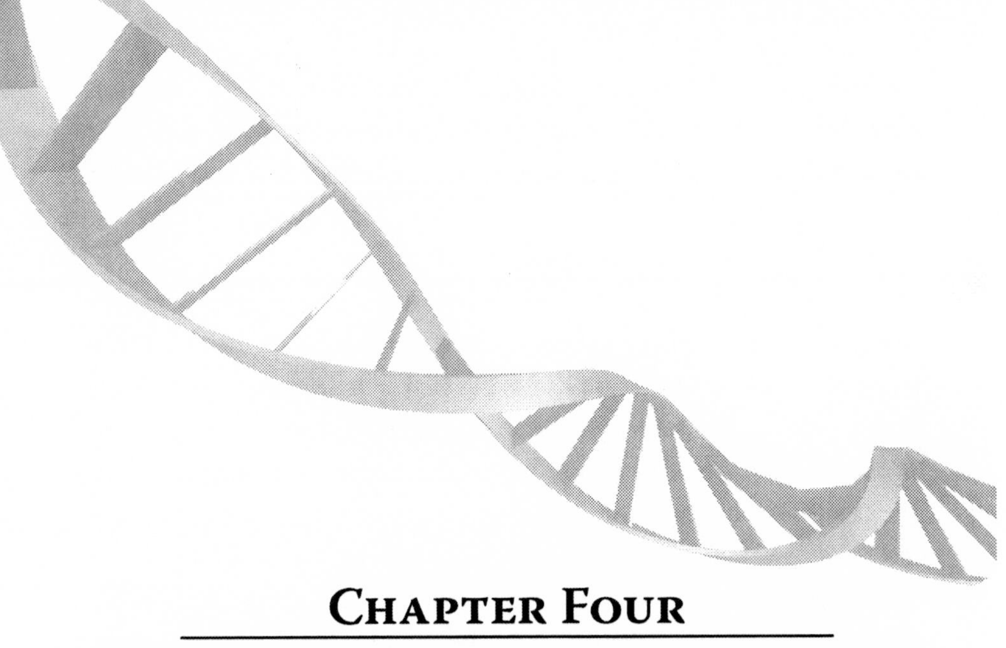

Chapter Four

Search for a Stem Cell Clinical Trial

After I was turned down by the two heart transplant centers, my cardiologist had exhausted all options for treatments that might be available. At this time, I was still suffering a great deal of chest pain that was believed to be caused by blockages in my heart's minor arteries. Pain was an indication that these arteries were not providing sufficient blood to my heart muscles. Additionally, because my heart was not pumping an adequate amount of blood to meet the needs of my body, I was in congestive heart failure, or CHF. This progressive disease leads to certain death. I knew that my doctor, who was very skilled, was limited to providing only treatments that were approved by the FDA. When I questioned him about advanced treatments, he told me he was aware of stem cell research being conducted at Duke University by former colleagues and would try to obtain additional information. Many practicing physicians do not have enough time to stay informed about all of the many advanced treatments being studied at research hospitals around the world. Practicing physicians are able to stay up to date on new developments by reading medical journals, meeting with drug company representatives, and attending conferences related to their specialties. However, the information they receive is primarily about medical procedures and drugs that are already approved for general use. It became apparent that I would need to take it upon myself to do the research necessary to find an advanced treatment that might help me.

My doctor's stem cell recollection turned out to be about studies being performed on animals in a laboratory. Although not helpful at the time, the information did lead me to performing an extensive Internet search for therapies that use stem cells for repair of damaged human hearts. I am so thankful for the existence of the Internet. The best possible source of information was no farther away than my computer. This was very important because I was an invalid and not able to drive to public and university libraries or to remain there for extended periods of research. Additionally, I realized that material available in the library might not be as current as that which can be found on the Internet because the Internet is updated daily.

The Internet is actually a network of networks, linking computers to computers sharing a standard procedure for regulating data between them. Each runs software to provide, or "serve," information and for people to access and view information. The Internet is the transport vehicle for information stored in files or documents on other computers. The Internet itself does not contain information. It is a slight misstatement to say a document was found *on* the Internet. It would be more correct to say it was found through or using the Internet to access information that was stored on another computer linked to the Internet.

The World Wide Web, or WWW, incorporates all of the Internet services just described, and more. Through the WWW, or simply the Web, you can retrieve documents; view images, animation, and video; listen to sound files; and view programs that run on practically any software in the world, providing that your computer has the hardware and software to do these things.

You need two essential types of computer programs for finding and retrieving information from the Internet. The first program you need is called a browser. Two popular browsers are Netscape and Microsoft's Internet Explorer. A browser is a computer program that resides in your computer and accesses and displays files and other data available on the Internet. The second program, called a search engine, resides on the Internet and retrieves documents, files, or data from databases in computers connected to the Internet. There are many search engines available to the user. The most popular include Google, Alta Vista, and Yahoo.

Information is found by typing key words into the search engine which will then search literally billions of computer files to find those files or Websites which contain the chosen words. Amazingly, this takes only a small fraction of a second to accomplish. A single word or a string of words can be entered. If a string of words is entered, the search engine will find only those Websites or files that contain the words as they have been entered.

I began my research using Google for references to "heart," a simple enough term. To my surprise, there were 137 million articles. Obviously, I could not read all of those, so I had to narrow my search a great deal. The phrase "heart therapy" narrowed my search to nearly seven thousand entries. I opened hundreds of links and found references to a variety of advanced therapies. Studies involving open-heart surgery had to be ignored because I was too weak to survive such an operation again. Fortunately, I found about one hundred references to stem cell therapy for the heart. It was important that I followed every link on every Webpage that I opened. This had to be done, because the initial page might not contain the information needed, but a link on the page might lead to information that was more pertinent. Links may be words highlighted in blue and underlined, or they may be "buttons" on which you can click your mouse. This "clickability" link is a feature that is unique and revolutionary about the Web.

I also researched the nature of stem cells themselves. I downloaded and printed many scientific papers and articles describing stem cells, including the various types that exist in the adult body, and how they are harvested. I found that most stem cell research was being done in laboratories on insect and animal subjects. The reported results were promising for therapies that might develop in the future, but they would do me little good if I didn't live long enough for the studies to reach clinical trials. My findings regarding stem cells are presented in Chapter Six. Since stem cell therapy appeared to be the most promising approach for repairing the damage to my heart, I needed to find studies on human subjects. Searching further, I was fortunate to find that clinical studies involving repair of damaged hearts using adult stem cells from patients' own bone marrow were being performed on people in clinical trials.

The next step in my plan was to call every research laboratory and research hospital involved in adult stem cell heart therapy. At each facility, I planned to ask for the cardiology department and talk with the principal investigator of the study being conducted. I would also try to get references to others in their field that I might not have discovered.

Before I could start making the calls, I was reading a *Fortune* magazine in a doctor's office and was amazed to come upon a short article about success with adult stem cell therapy in a clinical trial being conducted by cardiologist Dr. Douglas Losordo, MD, at the Caritas St. Elizabeth's Medical Center in Boston. As soon as I got home, I searched for Dr. Losordo's name on the Internet and came up with information about how to contact his research coordinator. Before making the call, however, I also did a thorough search on clinical trials and what is involved in participation in them. This is reported in Chapter Eight.

Chapter Five

Participation in an Adult Stem Cell Clinical Trial

After exhausting all possible therapies, my wife, Lynn, was beginning to feel that we should just accept the limitations on my life. The long and trying process to qualify for a replacement heart had taken a toll on both of us. Although we had experienced seven years of a joyous marriage before my near-fatal heart attack, stress during the last four years had been very hard on us. Nevertheless, I could not believe that I would have to live with increasing heart failure and premature death.

My condition, congestive heart failure (CHF), meant that my heart was becoming progressively weaker. Because it could not pump strongly, not enough blood went to the rest of my body, and my blood pressure fell. Backup of fluid in my lungs gave me shortness of breath, and my legs and ankles swelled alarmingly. My blood pressure was so low my kidneys could not work at full capacity to remove excess fluid. My future would consist of a declining ability to do things for myself and frequent hospitalizations to remove excess fluid. Additionally, I could not walk across a room without pain that required morphine for relief. Lynn was afraid to leave me alone while she was at work because I had fallen several times.

Unfortunately, I had been reduced from a vital, active professional to someone who awoke in the morning to eat, take a nap, have lunch, take a nap, have dinner, take a nap, then wake long enough to watch the late news and go back to bed. The impact on Lynn was also very severe. Stress was constant, and she became angry at the futility of the situation. Although she didn't want to incur additional expense for yet another doctor and more medications, she listened to the advice of family and friends and started seeing a psychiatrist whose counseling and antidepressants were a big help.

When alert, I concentrated as long as possible on using the Internet to search for clinical trials of advanced treatments that might ease or cure my condition. Had it not been for the EECP, described in Chapter Three, I doubt I would have even had enough strength to use the Internet.

I have always been an optimist, and after careful investigation, I am willing to try almost anything. However, my outlook was somewhat clouded by the amount of morphine I was taking. Luckily, Lynn is highly analytical and acted like a brakeman on a runaway train of possibilities. She asked the hard questions concerning safety, cost, availability, and effectiveness of each possible treatment. At times, these were not easy discussions.

During my Internet research, I had found that stem cell treatments for the heart might be available in Thailand.[6] Lynn became very cautious and could not share in my excitement. She was particularly worried about the logistics of a trip halfway around the world and a possible lack of controls. Then, after I found information about the clinical trial in Boston of a promising heart therapy using adult stem cells, she became more interested. I contacted the trial's research assistant, Daniela Grasso, and was then interviewed over the phone several times by her and Dr. K.K. Poh, a clinical research fellow participating in the study.

During the last week of December 2004, Lynn happened to answer the phone when Dr. Poh called to request my medical records. Dr. Poh had already conducted several screening calls with me, but this was the first time Lynn had a chance to speak with anyone on the trial team. Dr. Poh reviewed some of the same questions he had posed to me and was delighted to find that Lynn was also knowledgeable about my medical history. Then, she had questions for him about the study. For her own peace of mind, she needed to know the credentials of the hospital where the study was taking place, which authoritative medical body had sanctioned the study, and what was involved in arranging an evaluation. Dr. Poh assured her that this stem cell clinical trial was an FDA-monitored study.

Only then did Lynn become comfortable with sending off my medical records. Now, she was completely supportive. We once again worked as a team, doing everything possible to get included in the study.

Nevertheless, Lynn still had some doubts about my being physically able to travel back and forth to Boston. Because of the possibility of another heart attack, we had not traveled by air for four years. There were financial aspects of the trip to consider as well. In mid-January, the program coordinator, Marianne Kearney, sent a proposed schedule for us to come to Boston in early February for a visit with the principal investigator, cardiologist Douglas W. Losordo, MD. Lynn was unsure about paying for airplane tickets until after my medical records had been thoroughly evaluated and we knew with certainty that I would be considered as a candidate for the clinical trial. It was well that she was cautious. This was borne out when, after receiving my medical records including two prior angiograms, Dr. Poh requested a more recent angiogram as the latest was now over two years old. He explained that the study protocol required that two independent cardiologists review a recent angiogram and agree that no other treatment options would help me. We were able to get a new angiogram scheduled and completed within a week. The hospital immediately sent a CD with the angiogram films by FedEx.

On January 25, Marianne Kearney confirmed that both independent cardiologists had reviewed my angiogram and had agreed that there were no conventional treatments that could be of help to me. With this news, Lynn was eager to set up an appointment with Dr. Losordo. Marianne Kearney wanted to make it for February 2, but we needed at least fourteen days to get a reasonable price on airline tickets. Marianne Kearney and Lynn agreed on February 9. I was anxious to get to Boston as soon as possible. My concern was that the clinical trial might be full before we got there. Later, we found that just the opposite was the case. They were actually having a difficult time finding heart patients who had exhausted all other possible avenues of treatment *and* had found out about the clinical trial.

Although it appeared that I made an instantaneous decision to join a Phase I clinical trial, I was a little apprehensive about becoming one of only twenty-four patients in a trial that had the purpose of testing the safety of the procedure for the very first time. After a great deal of thinking during the dead of night, I realized that I had no alternative. Besides, if the stem cell study were a success, it would not only save my life, it would help save the lives of perhaps millions of other people in the future.

This aspect was appealing to me; I had saved two lives in the past, which had given me a great deal of personal satisfaction. The first was a homeless person who had fallen asleep behind some shrubbery at the First Presbyterian Church in Orlando, Florida, when the outside temperature was 29°F. I was a volunteer security person on "cold nights" when we took the homeless into Fellowship Hall to give them a warm place to sleep and a hot breakfast. During the night, I periodically patrolled the block surrounding the church property to invite in possible stragglers who had not found the entrance. Luckily, I spotted a lightly clothed man with my flashlight as he slept on the ground behind some bushes. I am certain he would not have awakened in the morning after a whole night of exposure.

The second time, Lynn and I were having dinner in a local restaurant when a woman at a table behind us suddenly shrieked,

"Does anyone know the Heimlich maneuver? My husband is choking to death!" When we spun around, we saw a man standing up, clutching his throat and trying desperately to breathe. No one else moved to help, so even though I had never done it before, I rushed behind the man, put my arms around him, and gave two quick thrusts into his solar plexus with my clenched fists. Immediately, he coughed up the food lodged in his throat and gasped for air. I cannot tell you how good that made me feel.

Lynn booked a flight from Orlando for Wednesday morning at 7:00 a.m. to arrive in Boston by 10:30. Not wanting Lynn to take an extra day off work to travel the night before, we were up at 4:00 to get out of the house by 5:00. Our luggage was already packed and by the door. If I was accepted into the stem cell trial, we had to pack for two and a half weeks of cold weather. Besides a big suitcase on wheels, we had a duffel bag on wheels, another bag containing my breathing machine for sleep apnea, Lynn's breathing machine for the same, along with hoses and masks. We also had a briefcase with books to read, papers related to the trip, and a small carry-on containing a three-week supply of the nineteen different medicines I was taking.

Knowing the trepidation Lynn felt, I gave her a kiss as we went out the door and vowed that if this didn't work out, I would accept life as it was. She could only laugh, as she knew I would never quit.

At the Orlando International Airport, Lynn found a wheelchair for me and checked our bags. Off we rolled, with Lynn pushing me through the main lobby, which was silent except for the sounds of shops just beginning to open. Suddenly, we realized we must now go through the security checkpoint. *How are we going to navigate this?* we wondered.

The security checkpoint was quite an experience. I was detoured for a personal check as I have a defibrillator/pacemaker implanted in my chest and cannot go through the regular scanner. Lynn filled several bins with my cane, belt, pocket change, shoes, our breathing machines, and Lynn's boots, purse, briefcase, cell phone, and watch. It was a challenge for Lynn to do this while watching to see if I would fall getting out of the wheelchair. When she tried to grab all of our stuff off the conveyor while watching out for me, she created a small traffic jam of early morning travelers.

After we got to the airside terminal, we relaxed a little and ate some breakfast. At the gate, the attendant rearranged our seats to be together and requested a wheelchair in Boston. Her helpfulness was the first of many pleasant experiences. Because I was in a wheelchair, the flight attendants made everything even easier by boarding us early and helping stow our bags.

In Boston, a skycap pushed my wheelchair through the long corridors from the gate to the baggage claim area in the terminal. Once Lynn located our luggage, we called the rental car company to send a van to take us to the off-site rental lot.

Driving from the airport to Caritas St. Elizabeth's Medical Center turned out to be easier than anticipated as we had only a couple of turns to make after we got off the Massachusetts Turnpike. When we arrived at the hospital, Lynn turned the car over to a valet and found a wheelchair for me.

I thought it would make a more favorable impression if I were able to get around without a wheelchair, so I insisted on leaving it in the hallway and walked into the Vascular Medicine Department using my cane. We let the receptionist know we were there and found that there was time for lunch in the cafeteria before our appointment. After eating, we were making our way back to the Vascular Medicine Department, when Marianne Kearney, the program coordinator, magically appeared, introduced herself, and inquired about our trip and accommodations. She quickly put us at ease and soon introduced us to Dr. Poh, a tall, quiet, dark-haired man in his late twenties who gave the impression of spending a great deal of time over a microscope. Dr. Poh, who was from Singapore, seemed to choose his words carefully, as if unsure of his excellent English. Then, we were led to a tiny exam room. Lynn took a chair. I sat on the exam table, and Marianne Kearney and Dr. Poh stood near the door. Dr. Poh took the folder Marianne Kearney had brought into the room and started taking my medical history once again. Lynn produced a list of my medicines and reviewed it with Dr. Poh. Then he listened to my chest and checked the pulses in my ankles and neck. We were soon joined by Dr. Losordo, a handsome, distinguished-looking man with intelligent eyes and a serious face that looked as if it could easily break into a smile. Dr. Losordo reviewed my history yet again and told us about the clinical trial. He then asked if we had read and had questions about the "Informed Consent Experimental Treatment Form" that had previously been sent to us. A copy of this form is attached as Appendix A. It is a good example of what you can expect to receive before entering a clinical trial.

Dr. Losordo queried us about our interest in going forward with testing for the stem cell clinical trial. I had a few questions for him dealing with other studies. I also wanted, and received, assurance that if I was one of those who received a placebo instead of my own stem cells, I would later receive the actual stem cells. Dr. Losordo assured me that if anyone's health seriously declined over six months, that person's number would be unblinded, and he or she would receive their stem cells at that time, if they had not already been received. Once all the questions had been answered, Dr. Losordo shook hands with us, and Marianne Kearney gave us our schedule for the rest of the day. First, blood was drawn, then I had an EKG, a urine sample was collected, and I had a chest x-ray. Every step of the way, we were escorted by either Marianne Kearney or her assistant, Daniela. They seemed to know the moment I finished with one test and reappeared to take us to the next. The last test of the day was a treadmill exercise test. This turned out to be a problem for everyone.

I was mentally and physically exhausted after starting the day at 4:00 a.m. and was barely able to stand and move onto the treadmill. Fortunately, an unexpected problem arose. My pacemaker was keeping my heartbeat at sixty beats per minute, so the treadmill team could not measure the effects of exercise on my heart rate. The treadmill test was an important criterion for entry into the study, so this was a very serious problem and disappointment registered on everyone's faces. After a quick call to Dr. Losordo by Marianne Kearney, Dr. Losordo and Dr.

Poh were in the room immediately. Following a huddle in the hall, they came back with good news. The treadmill test would be rescheduled for the next day, Thursday, so that a specialist from the pacemaker company, Guidant, could be present to turn off the pacemaker for the exercise test and then restart it after the test. The delay allowed me to get a much-needed full night's sleep. Although Lynn was only trailing along from test to test, she was also ready for some downtime. We retrieved our coats from the Vascular Medicine office and left the hospital for the day.

We were staying with my daughter, Kate, who lives in Boston, but had forgotten to get directions from the hospital to her apartment, so we only had directions downloaded from the Internet. It was already dusk as we left St. Elizabeth's. With me reading directions in the growing darkness and Lynn trying to see tiny and sometimes non-existent street signs, we soon realized we were lost. Just then, an anxious Kate called on Lynn's cell phone, and using her directions, we were on her doorstep in no time.

Kate suggested a neighborhood restaurant, and we were soon warm and well fed. After dinner, Kate and Lynn got me settled in and went to the grocery store to pick up supplies for my special diet. When they returned, we were ready for bed and a sound sleep. Lynn set the alarm to allow plenty of time for us to get back to the hospital at eight the next morning. Thankfully, our first big day was over!

Thursday was another long day, and we didn't realize just how important it would be for us. First, there was a nuclear stress test (SPECT) then a CAT scan of my lungs because yesterday's chest x-rays had shown a slight shadow. Luckily, my chest was clear because the shadow was only a scar from an old surgery.

While we were waiting in the Vascular Medicine Department for yet another blood test, a tall Texan named Robert Young, with a hardy voice and grip, saw us talking with Marianne Kearney and came over and introduced himself and his wife. One of the first people in the study (number three), he was back for his one-year, and final, checkup. Bob Young turned out to be the person mentioned in the *Fortune* magazine article I had found last year. His success with the stem cell therapy was remarkable. He had gone from being a total invalid like me to again being a robust and healthy rancher. What an inspiration to us!

After they left, my blood was drawn again, and we went to the cafeteria for lunch where once again, we got a chance to chat with Bob Young and his wife. They answered all of our questions about Mr. Young's health and his experiences in the trial. Of course, we were delighted that they had come over to chat.

After lunch, I had an echocardiogram, and then we went back to the treadmill. This time, the small double room for the treadmill test was packed with people. There was a desk inside the door on the right and another behind a narrow partition on the left. The room also included a bed, two treadmill machines with wires attached to overhead monitors, and two chairs. Altogether, the room could not have been bigger than eight feet by eight feet, and it was packed with the pacemaker technician, an exercise technician, the exercise technician from the day before, Marianne Kearney, Dr. Poh, Dr. Poh's intern, Dr. Losordo, and Lynn. The group continued to grow as the technician turned off my pacemaker and electrodes were attached to my chest to monitor my heart. Lynn feared that exercising without my defibrillator could cause my heart to stop, but we had talked about this at lunch and agreed that I should do my best, and if I had a problem, we would be in capable hands. Neither of us had a clue about how critical the treadmill test was to admission into the clinical trial. However, the gathering crowd testified to its importance! After a false start, I was soon walking slowly on the treadmill machine and gasping for air as the pace was increased at three and four minutes. I tried my best to continue but had to stop at four and a half minutes. Everyone was suddenly smiling! I was relieved just to have it safely over. My pacemaker was then reset, and we were done for the day.

I had Friday off while the team met to discuss my candidacy for admission into the clinical trial. Marianne Kearney told us that she would call after the meeting to let us know if I qualified. We were tired and unsure about the future, but we easily found our way back to Kate's apartment and after a quick dinner and some relaxation, were once again asleep early.

Friday morning we slept late and awoke to new snow covering the ground. We both loved the magical quality of falling snow, but just before lunch, our enchantment was interrupted by the ringing of the phone. It was a very happy Marianne Kearney. I was in!

Marianne Kearney had some menu suggestions because I now needed extra calcium from milk, yogurt, and spinach. On Saturday, I started injections of Neupogen that caused my stem cells to move from my bone marrow into my blood. The injections continued for each of the next five days. We arranged to meet Dr. Poh at 10:00 in the morning to begin the shots and to have blood drawn.

Saturday, Sunday, Monday, and Tuesday, I had more shots and blood drawn. On Tuesday, we were told that my blood cell volume looked good and that my stem cells would be collected on Wednesday by a process called apheresis. I would have a heart catheterization on Thursday so that Dr. Losordo could inject the stem cells into my heart muscle. We brought books to read, as it takes four to six hours to collect the stem cells. The stem cell collection took place in a dialysis unit using a special centrifugal machine. Although this big machine looked imposing, covered in tubing with large internal cylinders and a bag attached to the right side, the two nurses in charge explained it was specially developed by Baxter Healthcare to isolate the CD34-positive stem cells that were needed to form tiny new capillaries and stimulate muscle growth in the heart.

After a large catheter was inserted into the artery on the right side of my neck, the nurses made me comfortable in a bed and then hooked me up to the tubing. All of my blood was pumped slowly through the hoses into the machine, a pint at a time, for about twelve cycles. The machine then spun the blood and separated the lighter stem cells from heavier red blood cells and pushed the stem cells into a collection bag. The collected stem cells at this point were still mixed with light, creamy-colored white cells that were also captured.

After about six hours, the process was completed. I napped most of the time, waking only to eat scantily of a light lunch. Lynn had a large recliner in which to relax and read, but she spent much of the time talking with the nurses. Once the collection was completed, the bag with my precious stem cells was signed over to a lab technician. Because this was a double-blinded study, she was the only one on the team to know if I would receive my stem cells or a placebo (a saline solution). After she processed and extracted the stem cells, she called the FDA and was given the number nineteen, and I became nineteen in the study. She was then told to either freeze my stem cells and prepare the placebo or ready my stem cells for injection into my heart. The study would stay blinded until all twenty-four participants had finished the study. Like all the other families before us, we joked with her about paying for her upcoming wedding if she made sure I would be well enough to dance at her reception.

Thursday was the most important day for us. We arrived early and waited outside the heart catheterization lab for my turn, which was delayed by an emergency. Then, a nurse took me back to get started, and Lynn was sent off to the cardiology waiting room. A technician turned off my pacemaker, and nurses prepared me for surgery.

The heart catheterization required that a thin catheter be threaded in from a slit in my groin artery to the inside of my beating heart. Dr. Losordo then mapped the electrical activity of my heart using a three-dimensional mapping system, developed by the Cordis Unit of Johnson and Johnson, to find the working and nonworking areas of my heart muscle. Injections of the CD34-positive stem cells were made at the edges of the non-functioning areas of the heart to promote blood flow and bring the dead muscle back to life. Since my first research work involved development of weapons during the Cold War, I later explained to Lynn that the Cordis device was based on a missile-tracking system developed by Israel. We both laughed about how turning swords into plowshares might save my life.

The mapping and injection of stem cells was expected to take one to two hours, but Lynn started to fidget after one hour. About this time, Marianne Kearney appeared and told her that it would not be much longer, but my heart had stopped and had to be shocked to get it started again. Lynn would have been worried except that she had learned over the past four years that my fragile heart stopped every time I had a heart catheterization. The need for shocking my heart became routine to her, although during the first few days after my original heart attack, she was very concerned about my needing that life-saving procedure. After Dr. Losordo finished, he gave Lynn a three-dimensional map of my heart showing where the stem cells, or saline, were injected. When the procedure was over, I had to stay overnight at St. Elizabeth's and Lynn stayed the night at Kate's.

I was released from the hospital the next afternoon, Saturday, but I had to take it easy the next few days and go back for a checkup the following week. Sadly, Lynn had to fly back to Orlando Sunday evening to return to work. Kate looked after me the next few days and got me on the plane for home Tuesday after I had been checked and approved to fly.

The study required follow-up visits to monitor my health at one month, two months, three months, six months, nine months, and a year. In addition, Dr. Poh had to be notified and sent the results of any hospitalizations within forty-eight hours. The clinical trial was required to inform the FDA of each occurrence so that it could determine if there were problems caused by the stem cell experimental procedure resulting in injury or death.

The one-month visit in March involved a physical exam, EKG, echo, urinalysis, and blood tests all on the same day. We booked a flight leaving Orlando at 9:00 a.m., arrived in Boston at 11:24, and returned the same day. Marianne Kearney supplied us with taxi vouchers for travel from the airport to the hospital and back so that time was not wasted on renting and returning a car. The day went smoothly, but we were a little disappointed that we did not yet see any improvement in my health. Lynn was convinced that I had not received my stem cells, but she didn't say anything to me. However, I had been thinking the same thing, and I brought it up with the doctors. I had hoped for the same almost immediate improvement Mr. Young told us about when we spoke with him on our earlier visit. Dr. Poh assured us that it might take four to six months to see any significant change. The day went so smoothly that Lynn and I decided that I would return to Boston by myself for the two-month checkup in April. I still needed wheelchairs to get through the airports, but I took the same direct flights up to Boston and back. Kate met me at the airport, took me to the hospital, and stayed with me for my checkup. I still did not show any improvement, which convinced me that I had received a placebo instead of my own stem cells.

Thankfully, I got home without incident, but Marianne Kearney suggested that Lynn come for the next visit, as it entailed another treadmill exercise test along with a physical exam, EKG, echo, nuclear stress test, urinalysis, blood testing, and questionnaire.

Because the three-month visit in May took two days, we stayed with Kate again. We traveled much lighter this time, with only one carry-on bag, one suitcase, and a bag with our breathing machines. I was still unable to walk far enough to get through the airports and needed wheelchairs. This time, our mood was expectant, as we had started to see a few changes in my health. I was awake more now, able to prepare meals, and I took our dog out for short walks. I started to believe that maybe I did receive my stem cells after all. Lynn and I wanted the tests to confirm improvement one way or another. While the nuclear stress test didn't show any real change, I was easily able to go for another thirty seconds on the treadmill, and there was some improvement in my ejection fraction in the echo test. I had read that early stem cell studies on animal subjects suggested that heart-pumping capacity, as measured by the ejection fraction, might be boosted by between 5% and 30%. At this point, any improvement for me would be great as I started the study with an ejection fraction below 30%. A normal ejection fraction is between 50% and 60%. A simple improvement to 35% would give me more strength and endurance. Nevertheless, neither of us was truly convinced that I had received my stem cells. Dr. Poh again said it was too early to expect dramatic results. We went home more or less encouraged, even though we still had doubts.

Between the three-month visit in May and six-month visit in August, some amazing changes took place. Two particularly dramatic events demonstrate these changes. Neither was possible before the stem cell clinical trial.

The Christmas of 2004, before I was in the clinical trial, we drove up to Columbia, South Carolina to visit Lynn's older daughter, Cathy. Lynn and I stayed at a nearby hotel because I was unable to climb the stairs to the bedrooms of Cathy's home, and we stayed only two days because the trip completely wore me out. We made a second visit over the Fourth of July weekend when I had been in the clinical trial for five months. This time, I stayed wide-awake and alert during the entire seven-hour drive. When we arrived in mid-afternoon, Cathy and her family were floating in their backyard pool and invited us to join them. Lynn expected me to collapse into a chaise lounge to rest, but, to her surprise, I wanted to join them in the pool. It was the first time in over five years that I had had enough energy to do anything of this sort. Not only that, but I had no trouble going upstairs to unpack the swimsuits, and I was the last one out of the pool. We had a slow, relaxing weekend, and I was able to keep up with the others without needing any naps. When it came time to leave, Cathy and Lynn heard a light bump, bump, bump on the stairs. They both jumped up in alarm and rushed to the stairs to see me bringing down a heavy suitcase! As Lynn began to chastise me for lifting the suitcase, I reminded her that I had been able to manage the stairs several times every day. Even more amazing, after our long drive home, I suggested a swim in the pool at our apartment complex. We exercised in the pool daily for the rest of the summer. What a great change!

Two weeks later, we were on the road again for a two-and-a-half-hour trip to Tampa for the wedding of my oldest grandson. We checked into the hotel where my family was staying and sat in the lobby until after midnight, talking, laughing, and catching up with each other. I did not need a nap and was one of the last to go to bed. The next day, we met for breakfast and then went to the afternoon wedding and reception, which lasted long into the evening. Not only was I not exhausted, but Lynn and I spent a great deal of time on the dance floor. After breakfast the next day, we were off to visit my sister who was ill and could not make the wedding. We then drove back to Orlando, and I suggested we stop at the grocery store for our weekly shopping. Lynn and I could not believe the change in my energy level. I felt as well as before my heart attack over five years ago.

My six-month follow-up visit to St. Elizabeth's was in August. This two-day visit included a physical exam, EKG, echo, nuclear stress test, urinalysis, blood tests, and questionnaire. While we were both anxious to get a definitive answer as to whether or not I had received my stem cells, we believed that the immense improvements in my capabilities gave us an answer. When we met with the stem cell clinical trial team, we shared all of the changes I had experienced during the past month. Not only could I do mild exercises now with little or no pain, but I was also vacuuming the carpets, taking out the trash, and doing all of the other normal household chores. My improvement allowed me to work with my pain medicine doctor to reduce my morphine and give me a clearer head. I was steady on my feet and did not need a cane.

The stem cell study team told us that it was unlikely that improvement from a placebo effect would begin after four months and last through the sixth month. Additionally, results from the clinical tests showed improvement in my heart function. My ejection fraction (EF) had improved from 30% to 35%, meaning that my heart was pumping more blood to my body than it had been doing before the clinical trial began. This seemed to correlate with my exercise tolerance and general feeling of well being.

I went to Boston for an uneventful nine-month checkup in November and reported continuing improvement. I felt no adverse effects from the treatment of my heart with my own stem cells.

During the tenth and eleventh months, I was shocked by a general decline in my health. I became short of breath while performing routine tasks, and the pain in my chest was becoming increasingly frequent and more intense. I reported this worrisome condition to the clinical trial team by telephone and requested that my portion of the study be unblinded and said that I wanted to be given my stem cells if I had previously received a placebo. Dr. Losordo suggested that we wait until the twelve-month visit so that he could make a decision based upon a clinical examination.

Based upon examinations and tests during the twelve-month follow-up visit, Dr. Losordo requested, and was granted, permission from the FDA to unblind my portion of the study. There was genuine shock when the trial team told us that I had actually received the placebo instead of my stem cells! Why had I shown clinical improvement without the stem cells? The improvement in my ejection fraction and other measurements of heart function had not decreased. But why was my pain increasing? Why was I short of breath when performing routine activities?

Dr. Losordo had no definitive answers, but he offered the explanation that when my stem cells were harvested from my bone marrow by the granulocyte colony stimulating factor (Nuepogen), the stem cells circulated in my bloodstream for four days before they were harvested. During that time, the stem cells reached the damaged part of my heart muscle and stimulated the heart to begin repairing itself. However, not enough cells reached my heart to affect a permanent repair.

Two weeks later, the clinical trial team harvested my stem cells again, and Dr. Losordo injected them into my heart. This time, I was sure that I actually received my stem cells. And I had no doubt that the stem cells would do a complete and effective job of repair, as had happened to other participants who actually received their stem cells. This turned out to be true for me also. In the fifth month after receiving my stem cells, I again regained my health.

While Lynn and I were focused solely on my own improvement during the clinical trial, the primary goal of the trial team was to demonstrate that the procedure was safe and that there are no serious side effects. The team's next trial phases, which have already started, will be oriented toward demonstrating overall effectiveness and the most effective dosage.

As far as Lynn and I are concerned, the stem cell therapy has been a complete success. We have our old lives back again and look forward to a long and happy future.

Chapter Six

Stem Cells

My education about adult stem cells and their almost miraculous curing powers came after a massive heart attack that severely damaged the muscles of my left ventricular chamber, which is the main pumping chamber of the heart. The muscle damage turned me into an invalid. I suffered acute pain and was diagnosed with congestive failure, which is progressive and results in premature death. Over a period of four years, my cardiologist, Dr. Mark Milunski, treated me aggressively with all available conventional drugs and therapies, but they were not effective. During that time, I was also evaluated at two heart transplant centers, but because of my age and other factors, I was not eligible for a heart transplant.

Our doctors do everything within their power to ease our suffering, but they are limited to therapies that the Food and Drug Administration (FDA) has approved for general use. Surprisingly, physicians are not likely to know about all of the advanced treatments that are in clinical trials in hospitals and laboratories around the world. Although they keep up with approved new therapies by reading medical journals and participating in required continuing education courses, their busy practices keep them from spending the large amount of time necessary to be aware of all of the advanced treatments not yet approved for general use.

With nowhere else to turn, I decided to learn all I could about advanced treatments that could benefit me. I performed an exhaustive search on the Internet for reports on treatments already in the clinical trial phase but not yet available to practicing physicians. The search led to a clinical trial using adult stem cells from one's own body to repair muscular damage to the heart. I was successful in becoming a participant in a trial in which stem cells were coaxed out of my bone marrow using a special biologic agent. Dr. Douglas Losordo, the principal investigator at Caritas St. Elizabeth's Medical Center, Tufts University School of Medicine, then injected my stem cells into my heart muscle at the very edge of the damaged area. After about five months, I noticed a huge improvement in my physical capabilities, and my pain was nearly gone. The stem cells had improved blood flow to my heart muscle and may have started growth of new heart muscle. With improved heart pumping capacity, I went from being a dying invalid to a happy and physically capable human being. Chapter Five presents the details about my participation in this pioneering clinical trial.

To understand the nature of stem cells, I collected many scientific papers and articles describing stem cells of various types. Before the literature search, I had some idea about the nature of stem cells, but my knowledge was very limited, and I was subject to some misconceptions.

The popular media had led me to believe that all stem cells are controversial embryonic cells. Fortunately, that is not the real situation. What I found was that medical scientists and research physicians are working in hospitals around the world to utilize umbilical cord blood stem cells and adult stem cells from patients' own bodies to treat many diseases.

What exactly is a stem cell? When I researched the subject of stem cells on the Internet, I came away with an entirely new understanding. I am not a physician or a biological scientist so I can give a simple, easy-to-understand, non-technical explanation. A stem cell is a cell in the body that has two unique characteristics. First, a single stem cell can divide and make another stem cell just like itself. It can do this repeatedly, yielding many stem cells. The second unique feature of a stem cell is that it doesn't know what kind of body cell it will become. The stem cell may become a heart cell, a liver cell, a skin cell, or whatever. Scientists say that a stem cell is "undifferentiated."[7,8,9,10]

When the doctors removed stem cells from my bone marrow, the first thing they did was to make them copy themselves in the laboratory to obtain a quantity sufficient for repair of my heart. When they had made enough identical stem cells, they used a catheter to inject them into my heart muscle. When the stem cells were exposed to my heart cells, they realized that they should also be heart cells. Scientists do not yet understand how this change is triggered, but they do know that it happens.

Adult Stem Cells

Bone marrow, the tissue comprising the center of large bones, is currently the most abundant source of adult stem cells. It is the place where blood cells, including stem cells, are produced. Two types of stem cells are produced in bone marrow: hemopoietic, which can produce blood cells, and stromal, which can produce fat, cartilage, and bone. Researchers have discovered adult stem cells in other parts of the body, including the heart, liver, uterus, fat, baby teeth, and many other organs and tissue.

In addition to adult stem cells, there are three other major types. There are embryonic stem cells, fetal blood cells, and cord blood stem cells.

Embryonic Stem Cells

Embryonic stem cells are derived from a group of cells in the early (four- to five-day) growth period of an embryo. With current technology, the embryo must be destroyed in order to harvest the stem cells within. This has caused a major nationwide controversy by those who consider an embryo to be a viable human whose destruction involves taking a life. President Bush has put severe restrictions on government funding of research using embryonic stem cells, although some states are taking it upon themselves to provide funding. Private organizations are also pouring hundreds of millions of dollars into embryonic stem cell research. Some scientists believe that embryonic stem cells are superior to others because they have a greater capability to become any kind of body tissue and are less subject to rejection when transplanted into humans. Other scientists do not agree, having shown that adult stem cells may be just as likely to become any type of tissue. To ease the controversy surrounding the use of embryonic stem cells, one scientist has reported an apparently successful method for removing stem cells from an embryo without destroying the embryo. However, this may or may not satisfy those individuals concerned with the use of embryonic stem cells for medical purposes. There is much laboratory research underway using embryonic stem cells. However, none of this research has led to clinical trials in humans.

Fetal Stem Cells

Fetal blood cells are among those least under study because of the difficulty of obtaining fetuses for research. In addition, there appears to be a disadvantage in the use of fetal stem cells because of the possibility of rejection upon transplantation.

Umbilical Cord Stem Cells

Doctors collect cord blood stem cells from umbilical cord blood. The blood is frozen in liquid nitrogen for later extraction of the stem cells. Dr. Robert Sears says that researchers are currently using cord blood cells to treat over seventy-five cancer and blood disorders. Dr. Sears also reports that survival rates are more than double when a person's own cord blood or a family member's cord blood is used compared to using an unrelated donor sample. There are several public and private frozen cord blood stem cell depositories nationwide, and some physicians are encouraging parents to have their baby's umbilical cord blood preserved for possible future use to treat serious disorders if they arise.[3]

Adult stem cells are currently the most widely used stem cells in clinical trials to treat disorders of many organs such as the heart and brain.

Adult stem cells have actually been used therapeutically in the United States since 1968. This means actual treatment, not trial or theory. A private company, Corcell, on its Website, lists almost eighty therapies using adult stem cells. Most early therapies involved entire bone marrow transplants that were successful because the marrow contained adult stem cells. Additionally, there are more than 300 clinical trials underway in the United States that use only adult stem cells to treat a variety of diseases.[11]

Chapter Seven

Stem Cell Storage

Because of the great potential of adult stem cell therapies, storage of adult stem cells is becoming a minor industry. Umbilical cord blood storage has been going on for some time. Universally applicable adult stem cells and stem cells from patients' own bodies are also being stored. Adult stem cell transplants have been used since the 1960s to treat a variety of diseases. In 1988 umbilical cord blood stem cells were used for the first time in bone marrow stem cell transplantation. Since that time, approximately 6,500 transplants have been conducted using cord blood cells worldwide as an alternative to embryonic or adult stem cells extracted from bone marrow. Cord blood and other adult stem cells show great promise for potential future applications including Alzheimer's disease, cardiac disease, muscular dystrophy, Parkinson's disease, and many others.

Cord blood banking is the process of collecting blood from a newborn's umbilical cord immediately after delivery and cryogenically storing it for future medical use if stem cell therapy should become needed by the donor or the donor's family.

Collection comes after the cord has been clamped and cut, so there is no risk or pain to either the mother or infant.

Thousands of families are banking their newborn's cord blood because the stem cells it contains are genetically related to their families and risks of future transplant rejections are minimized. If needed later, stem cells harvested from the umbilical cord blood can be developed into specialized organs or tissues to treat life-threatening diseases.

Cord blood stem cell research in clinical trials is already a reality. Cord blood stem cells are an early form of adult stem cells. Many researchers believe them to be superior to embryonic stem cells that have yet to be tested on humans in the United States. Because umbilical cord cells are obtained after the baby is born, they are free of the political controversy surrounding embryonic stem cells. While embryonic stem cells have not been tested, adult stem cells, including umbilical cord blood cells, have been successfully used to treat over seventy conditions, including Crohn's disease, brain and other cancers, sickle cell anemia, stroke, gangrene, corneal regeneration, spinal cord injury, juvenile diabetes, and heart disease.

Another approach to having adult stem cells readily available to fight the onset of disease and injury is the development by Osiris Therapeutics, Inc. of a universal adult stem cell drug called Provocel. The drug may be unique in that it may not require patient-specific matching. Using Osiris's approach, stem cells are mass reproduced in Baltimore, frozen, and then shipped to hospitals around the country where they are stored until needed. It is believed by the company that having the stem cells readily available will provide an opportunity to treat patients in acute settings. Provocel is a formulation of adult stem cells being developed to repair damaged heart tissue. It is given to patients through a standard IV line. In pre-clinical studies, Osiris reported that the cells responded to chemical signals given by the heart following injury. These signals attracted the cells specifically to the area of injury, where they participated in repair.

The company's current focus is also on the use of adult stem cells to improve outcomes in bone marrow recipients under treatment for Crohn's disease, to repair damage following a heart attack or congestive heart failure, and to prevent or treat arthritis.

Another cutting-edge company in the fast-growing stem cell industry is Neuronyx of Malvern, Pennsylvania, near Philadelphia. Neuronyx harvests adult stem cells from single donors. Its technology then permits replication of the cells to produce six billion new patient doses from each original donor. The company believes it is the first to inject adult bone-marrow-derived cells directly into heart attack patients. In Neuronyx's current patient study, the cells are applied with a catheter into the area around damaged heart muscle thirty to sixty days after a heart attack. The purpose is to stabilize and restore damaged tissue and prevent or reduce scar formation and progression to congestive heart failure. The company reports that the cells secrete a potent mixture of pre-regenerative factors such as cytokines or proteins that go directly to the area of injury and amplify or boost the host cell's own regenerative capabilities. This discovery is reported to have been made in spinal cord injury studies when improvements were noted in motor function in rats.

Neostem, Inc. is a New York-based company specializing in the collection, processing, and long-term storage of adult stem cells for autologous use (use in the same patient's body). The company's objective is to provide adult stem cell banking services for adults who wish to store their own stem cells for future therapeutic use in the treatment of diabetes, heart disease, and other critical health problems. The service is intended for health conscious individuals, individuals with family histories of heart disease or cancer, individuals diagnosed with chronic cancers, individuals who are exposed to radiation or harmful toxins in the workplace, and first responders (firemen, policemen, military personnel, etc., who may be exposed to lethal doses of radiation).

These adult stem cell harvesting and storage approaches are not presently covered by most insurance plans or Medicare. However, for those individuals with the financial means for storing their own stem cells, it is a viable way to protect their future health and give peace of mind that adult stem cells will be available if needed in the future to combat any number of serious diseases or injuries.

Chapter Eight

Stem Cell Clinical Trials

Practicing physicians do all they can to cure us when we are ill, but they are restricted to providing treatments that the U.S. Food and Drug Administration (FDA) has approved. Often, approval of advanced treatments comes only after many years of rigorous testing in the laboratory, on animals, and in a series of clinical trials on humans. There are miraculous stem cell treatments in human clinical trials right now, which could ease your suffering or save your life. If you or a loved one have exhausted all of the treatments offered by your physician, participation in a clinical trial may provide relief.

Today, we must all play a more active role in our own healthcare. Participation in clinical trials not only contributes to medical research, it also allows access to new research treatments before they are widely available.

In the United States, clinical trials are the only way to receive advanced treatments not yet approved by the FDA for general use.[30] Clinical trials on humans usually follow laboratory and animal studies.

Clinical trials, also known as clinical studies, test potential treatments in human volunteers to see if they should be approved for wider use in the general population. A treatment could be a drug, medical procedure, device, stem cell transplant, or other therapies. Before they are available to humans, scientists determine possible effectiveness or toxicity on laboratory samples and on laboratory animals. Scientists then advance treatments showing promise and acceptable safety profiles into human clinical trials.

Although "new" may imply "better," it is not known whether the potential medical treatment offers benefits to patients until clinical research on that treatment is complete. Clinical trials are an integral part of new treatment discovery and development and are required by the FDA before physicians can offer them in practice. The FDA protects participants of clinical trials and is one of the government agencies that provide information to those interested in participating.

Although efforts are made to control risks to clinical trial participants, some risk may be unavoidable because of the uncertainty inherent in clinical research involving new medical products. It is important, therefore, that people make their decision to participate in a clinical trial only after they have a full understanding of the entire process and the risks that may be involved.

Benefits of participating in a clinical trial include the following:

- Participants have access to promising new stem cell and other therapies that are not available outside the clinical trial setting.
- The therapy under study may be more effective than a standard approach.
- Participants receive regular and careful medical attention from a research team that includes doctors and other health professionals.
- Participants may be the first to benefit from a new therapy under study.
- Results from the study may help others in the future.

Possible risks of participating in a clinical study include the following:

- New therapies may not be better than standard care.
- New therapies may have side effects or risks that doctors did not expect.
- Participants in randomized trials will not be able to choose whether they receive the test therapy or a control substance (placebo).
- Health insurance and managed care providers may not cover any or all patient care costs in a study.
- Participants may be required to make more visits to the doctor than they would if they were not in a clinical trial.

People participate in clinical trials for different reasons. Some volunteer because they want to enhance medical knowledge. Others have tried all available treatments for their condition without success.

In a spring 2000 Harris poll of cancer clinical trial participants, 76% of the respondents said they participated because they believed the trial offered the best quality of care for their disease. Helping other people and receiving more and better attention for their own specific disease were other reasons cited.

Part of the reason I decided to participate in a clinical trial, involving transplantation of my own stem cells into my heart, was that I believed that what was learned would help other people with advanced heart disease. Others enter clinical trials in hopes of finding a potential treatment after traditional therapies fail. The major reason I went ahead with adult stem cell heart therapy was that I was told I had no other options.

Many clinical trials are free, except perhaps for travel, lodging, and other expenses. Some trials, usually conducted locally, may pay patients for their participation. People should not be tempted, however, to enroll in a clinical trial simply because a trial sponsor is offering a free potential treatment or because of the promise of money. People lured by compensation may overlook the known risks, may overlook a satisfactory conventional treatment offered by their physicians, or may fail to appreciate the potential for discovery of serious new side effects during clinical testing of a new treatment. Clinical tests are usually not a means for patients to receive long-term treatment for chronic diseases, although stem cell therapies may offer permanent improvements.

It is important to test stem cell and other medical therapies in the particular people they are meant to help. Trial guidelines, or eligibility requirements, developed by researchers usually include criteria for age, sex, type and stage of disease, previous treatment history, and other medical conditions. Some trials involve people with a particular illness or condition, while others seek healthy volunteers. Inclusion or exclusion criteria (medical standards used to determine whether a person may or may not be allowed to enter a clinical trial) are used to identify appropriate participants and to exclude those who may be put at risk by participating in a trial.

Scientists design every clinical trial to answer certain research questions. A trial plan called a "protocol" maps out what study procedures will be done, by whom, and why. Clinical trials often test therapies to see how they compare to standard treatments or to no treatment at all. Clinical trial teams include doctors, nurses, and other healthcare professionals. The team checks the health of the participant at the beginning of the trial and assesses whether that person is eligible to participate. Those found by the clinical trial team to be eligible, and who agree to participate, are given specific instructions and are then monitored and carefully assessed during the trial and after it is completed.

If you apply for a stem cell clinical trial, but the doctor or trial coordinator says you are not eligible, do not feel rejected. Clinical trials are planned very carefully to answer certain questions. Part of the process involves enrolling patients who are alike in certain ways. For example, a trial may be designed to answer questions about treating patients who have a particular stage of breast cancer or who have already received a certain type of chemotherapy. In order for the results to make sense at the end of a clinical trial, the team will enroll only those patients who meet the same criteria.

Another consideration is patient safety. For example, a new drug may be safe only in people with normal kidney or liver function. Therefore, the clinical trial team would not accept people with poorly functioning kidneys or livers into that study. In addition, people with a high risk for a therapy's side effects are ineligible.

Clinical trials are conducted in phases. In Phase I, researchers test an experimental treatment on a small group of people, generally twenty to eighty, to evaluate overall safety, determine safe dosage ranges, and identify side effects. This is sometimes called the mortality phase. Phase II trials of 100 to 300 participants are conducted to study the effectiveness of the experimental treatment and further evaluate safety. In Phase III trials, the treatment is tested on large groups of people, numbering 1,000 to 3,000, to confirm effectiveness, monitor side effects, and compare effectiveness to commonly used treatments. Sometimes Phase IV trials are conducted after a product or medical procedure is already approved for general use to find out more about the treatment's long-term risks, benefits, and optimal use, or to test the product in different populations of people, such as children.

Just because a certain trial may not be right for you does not mean that you should quit looking. Ask the trial doctors or the program coordinator if they know of other trials that may provide therapies for your problem. While you may not meet the inclusion criteria for a Phase I or Phase II trial, you may be accepted into a later trial phase.

Early clinical trials generally involve a "control" standard. In many studies, one group of volunteers will be given an experimental or "test" drug or treatment, while a control group is given either a standard treatment for the illness or an inactive pill, liquid, or powder that has no treatment value (placebo). This control group forms a basis for comparison for assessing the effects of the test treatment. In some cases, medical authorities consider it unethical to use placebos, particularly if an effective treatment is available. Withholding treatment, even for a short time, would subject research participants to unreasonable risks.

Volunteering for a clinical trial is no guarantee of acceptance. Similarly, there is no guarantee that an individual in a clinical trial will actually receive the stem cell therapy under study.

Whether a clinical trial participant receives the test therapy, a standard treatment, or a placebo is often decided by the trial team using a process called "randomization." This process can be compared to a coin toss that is done by a computer. During clinical trials, no one is likely to know which therapy is better, and randomization assures that treatment selection will be free of any preference a physician may have. Randomization increases the likelihood that the groups of people receiving the test therapy or the controls are comparable at the start of the trial, enabling comparisons in health status between groups of patients who participated in the trial.

The randomized control clinical trial is the standard scientific method for evaluation of stem cell treatment, pharmaceuticals, medical procedures, and other therapies. Randomized trials have been successfully used in both therapeutic and disease prevention trial studies.

In conjunction with randomization, a feature known as "blinding" helps ensure that bias doesn't distort the conduct of the trial or the interpretation of the results. Single-blinding means the participant does not know whether he or she is receiving the experimental drug, an established procedure for that disease, or a placebo. In a single-blinded trial, the research team knows what the patient is receiving.

A double-blinded trial means that neither the participant nor the research team knows during the trial which participants receive the treatment under trial. The patient will usually find out what he or she received at a pre-specified time in the trial.

My adult stem cell therapy trial was double-blinded. After a year and a half, my heart appears healthy and my physical ability is greatly improved, but I was not informed as to whether I received my stem cells or a placebo until my health started to decline.

Some treatments can have unpleasant, or even serious, side effects. Often these are temporary and end when the treatment is stopped.

Others, however, can be permanent. Some side effects may appear during the treatment, and others may not show up until after the study is over. Risks depend on the treatment being studied and the health of the people participating in the trial. All known risks must be fully explained by the researchers before the trial begins. If new risk information becomes available during the trial, participants must be informed.

Research with people is conducted according to strict scientific and ethical principles. Every clinical trial has a protocol that describes what will be done in the study, how it will be conducted, and why each part of the study is necessary. The same protocol is used by every doctor or research center taking part in the trial.

Clinical trials are federally regulated, with built-in safeguards to protect participants. The FDA has authority over clinical trials for drug and medical therapies and procedures regulated by the agency. This authority includes studies funded by government agencies as well as studies funded by industry or by private parties. An Institutional Review Board (IRB) must review and approve clinical trials. The board, which may include doctors, researchers, community leaders, and other members of the community, reviews the protocol to make sure the clinical trial is conducted fairly and participants are not likely to be harmed. The IRB also decides how often to review the trial once it has begun. Based on this information, the IRB decides whether the clinical trial should continue as initially planned and, if not, what changes should be made. An IRB can also stop a clinical trial if the researcher is not following the protocol or if the trial appears to be causing unexpected harm to the participants. An IRB can also stop a clinical trial if there is clear evidence that the new intervention is highly effective, in order to make it widely available. Additionally, to help protect the rights and welfare of volunteers and verify the quality and integrity of data submitted for review, the FDA performs inspections of clinical trial study sites and anyone involved in the research.

The FDA requires that participants be given complete information about the study. This process is known as "informed consent," and it must be in writing and signed by each participant. The informed consent process provides an opportunity for the researcher and patient to exchange information and ask questions. Patients invited to enter a trial are not obligated to join but can consent to participate if they find the potential risks and benefits are acceptable. A consent form must be signed by the participant prior to enrollment and before any study procedures can be performed.

Participants in a clinical trial work with a research team. Team members may include doctors, nurses, social workers, dietitians, and other health professionals. The healthcare team provides care, monitors participants' health, and offers instructions about the study. To ensure that the trial results are as reliable as possible, it is important for participants to follow the research team's instructions. The instructions may include keepings logs or answering questionnaires. The research team may continue to contact participants after the trial ends.

After a clinical trial is completed, researchers look carefully at the data collected during the trial before making decisions about the meaning of the findings and possible further testing. After a Phase I or II trial, researchers decide whether to move on to the next phase or to stop testing the therapy because it is not safe or effective. When a Phase III trial is completed, researchers look at the data and decide whether the results have medical importance.

The results of clinical trials are often published in peer-reviewed scientific journals. Peer review is a process by which experts review the report before it is published to make sure the analysis and discussions are sound. If the results are particularly important, they may be featured by the media and discussed at scientific meetings and by patient advocacy groups before they are published. Once a new approach has been proven safe and effective in clinical trials, it may become standard practice.

People can locate the published results of a study by searching for the study's official name or protocol ID number in the National

Library of Medicine's PubMed database. PubMed is an easy-to-use search tool on the Internet for finding journal articles in the health and medical sciences.

Participants have the right to leave the study at any time. At the same time, people need to know that circumstances may arise under which their participation may be terminated by the researcher without their consent. For example, sometimes it becomes evident early on that a trial is not working, and researchers know they are not going to get enough meaningful information to make continuation worthwhile. In addition, if an unexpected change occurs in the health status of a participant, perhaps toxic effects such as sudden kidney problems that may have developed, it would not be in the best interest of the patient to continue and certainly not consistent with the result the investigator is trying to achieve.

While it is true that stem cell clinical trials offer no guarantees, when standard treatments fail or none exist, clinical trials can offer hope. Participating in a clinical trial is one of the best ways to guarantee good care. This is especially true when no other course of treatment is available. People can reduce the confusion and uncertainty that often comes with deciding on whether to participate in a clinical trial by obtaining all the information available on various Websites; through phone calls; within FDA, HHS, and NIH offices; and from patient advocacy groups. It is important to know the different types of clinical trials, which questions to ask, and your rights as a participant. Find out what the risks may be and determine what level of risk you are willing to accept before you agree to enroll in a clinical trial for medical research.

As you consider enrolling in a clinical trial, you will face the critical issue of how to cover the costs of care. Even if you have health insurance, your coverage may not include all of the patient care costs. This is because some health plans define clinical trials as "experimental" or "investigational" procedures.

A growing number of states have passed legislation or instituted special agreements requiring health plans to pay the cost of the routine medical care received by a participant in a clinical trial. At the time of this writing, the following states require insurance companies to provide coverage for routine patient care costs:[31] Arizona, California, Connecticut, Delaware, Georgia, Louisiana, Maine, Maryland, Massachusetts, Michigan, Missouri, Nevada, New Hampshire, New Jersey, New Mexico, North Carolina, Ohio, Rhode Island, Tennessee, Vermont, Virginia, and West Virginia. You should contact your own state's insurance commissioner or your healthcare provider to determine coverage in your state at the time you are considering participation in a clinical trial.

In 2000, Medicare began covering beneficiaries' patient care costs in clinical trials. Up-to-date information about what Medicare will cover can be found on the Internet site of the Centers for Medicare & Medicaid.

There are two types of costs associated with a stem cell clinical trial: patient care costs and research costs. Patient care costs usually fall into two categories. The first is routine care costs, such as doctor visits, hospital stays, clinical laboratory tests, x-rays, etc., which occur whether you are participating in a trial or receiving standard treatment. These costs are usually covered by a third-party health plan, such as Medicare or private insurance.

Extra care costs associated with clinical trial participation, such as additional tests, may or may not be fully covered by the clinical trial sponsor and/or research institution. The sponsor and the participant's health plan need to resolve coverage of these costs for particular trials.

Research costs are those associated with conducting the trial, such as data collection and management, research physician and nurse time, analysis of results, and tests performed purely for research purposes. Such costs are usually covered by the sponsoring organization, such as the National Institutes of Health (NIH) or a private company sponsor.

Health insurance companies and managed care companies decide which healthcare services they will pay for by developing a coverage policy regarding the specific services. In general, the most important factor determining whether something is covered is a health plan's judgment as to whether the service is established or investigational. Health plans usually designate a service as established if there is a certain amount of scientific data to show that it is safe and effective. If the health plan does not think that such data exist in sufficient quantity, the plan may label the service as investigational.

Healthcare services delivered within the setting of a clinical trial are very often categorized as investigational and not covered. This is because the health plan thinks the major reason to perform the clinical trial is that there is not enough data to establish the safety and effectiveness of the service being studied. Thus, for some health plans, any mention of the fact that the patient is involved in a clinical trial results in a denial of payment.

Your health plan may define specific criteria that a trial must meet before extending coverage, such as sponsorship of the trial. Some plans may only cover costs of trials by organizations whose review and oversight of the trial are careful and scientifically rigorous, according to standards set by the health plan. Some plans may cover patient costs only for clinical trials they judge to be "medically necessary" on a case-by-case basis. Trial phase may also affect coverage; for example, while a plan may be willing to cover costs associated with Phase III trials, which include treatments that have already been successful with a certain number of people, the plan may require documentation of effectiveness before covering a Phase I or Phase II trial.

Choosing to participate in a clinical trial is an important personal decision. After identifying some trial options, it is essential that you consult with your personal physician. You should also consult with your friends and family. Your decision will likely involve the family because of the time and expense involved.

Clinical trials are being conducted to cure diseases of most parts of the body. The following are presented as examples of trials currently underway. Your research will reveal many more that are applicable to your own condition.

Heart

Stem cells have been applied in at least two different heart procedures. The first procedure is growing new arteries to rebuild heart muscle by providing much-needed blood to damaged areas of the heart muscle. The adult stem cell type used in this procedure is called the CD34-positive cell. There is also some speculation that this same cell induces actual growth of new muscle, but this process is not yet well understood. Another type of adult stem cell, the mesenchymal cell, is utilized to grow muscle and other tissue. Repair of heart muscle by mesenchymal stem cells appears to be most effective if treatment is given within ten days of a heart attack (myocardial infarction).

At Caritas St. Elizabeth's Medical Center, Tufts University School of Medicine, in Boston, twenty-four patients were treated in a clinical trial using adult stem cells to repair hearts damaged by myocardial infarction (heart attack). Eighteen patients were treated with their own adult stem cells, and a control group of six was treated with a placebo. The author was a patient in this study. Although the results of the trial have not yet been published, the author enjoyed remarkable recovery and is living a normal life.

At the Hanover Medical School in Germany,[12] thirty heart attack patients were injected with stem cells extracted from their own bone marrow, and an additional thirty control patients received placebos. Those who received the stem cells had a 6 to 7% improvement in heart function, while those with the placebo had none. Improvement of 6 to 7% may not appear significant, but it makes a huge difference in the quality of life of the patient. At the University of Düsseldorf in Germany, thirty-four heart attack patients were injected with their own adult stem cells and, again, were compared with a control group. Fifteen months later, the injected hearts were measurably improved, and the patients were generally healthier than their untreated counterparts. "Even patients with the most seriously damaged hearts can be treated with their own stem cells instead of waiting and hoping for a transplant," the chief researcher told *Die Zeitung* newspaper.[13]

In another controlled heart therapy study, a Brazil-Texas medical collaboration injected adult stem cells derived from bone marrow into patients' hearts.[14] They found that among fourteen patients, electromechanical mapping of the treated heart muscle revealed significant mechanical improvements of the injected heart segments after four months.

In all cases, a catheter was used to inject the stem cells through an artery and into the heart. Open-heart surgery was not necessary, minimizing risk and speeding recovery.

In another clinical trial in Brazil, scientists are preparing to launch a three-year clinical trial in which adult stem cells will be used to combat heart disease.[15] The study will involve 1,200 patients and will be coordinated by the National Heart Institute of Laranjeiras. The plan is to take samples of adult stem cells from the patients' own bone marrow and to inject these into their damaged heart tissue, invigorating damaged muscles and enhancing cardiac output. A major advantage of the treatment protocol is that the injected stem cells derive from the patient so there is no risk of immuno-rejection.

Suphatchai Chaithiraphan, chairperson of Chao Phya Hospital in Bangkok, Thailand, reports successful experimental treatment of twenty-seven patients whose own adult stem cells were coaxed from their bone marrow and injected via a catheter into their hearts. "With stem cell therapy, people who have not had access to heart transplants or resources to go to the hospital on a regular basis can be helped," said Kitipan V. Arom, chief cardio-thoracic surgeon at the Bangkok Heart Hospital.[16]

Vascular

Angiogenesis (the growth of new blood vessels) is perhaps one of the best examples of the wonders of adult stem cell therapy.[17] Three Japanese universities worked with forty-five men suffering severe leg circulatory problems. Two-thirds of the patients had diabetes, which makes healing very difficult; almost half had already had bypass operations in their legs, and almost half suffered gangrene. Many had sores that would not heal and experienced severe leg pain. All the subjects had their own stem cells inserted into one leg with the other leg receiving a placebo. Amazingly, all participants showed improvements in the stem cell-receiving leg. Toe amputations were avoided in three-fourths of the patients, while unhealed wounds improved in six of the patients who had them.

Kidney

Scientists at New York Medical College have developed a new procedure using adult stem cells to repair damaged kidney tissue.[28] The team, led by Dr. Maria Arriero, first extracted adult stem cells from samples of muscle tissue taken from healthy mice. The cells were then cultured *in vitro* and coaxed to develop endothelial tissue, which is a thin layer of cells that line blood vessels, lymph nodes, etc. When implanted into mice with damaged kidneys, the cells were observed to engraft, form new blood vessels, and substantially improve kidney function. The research not only demonstrates the potential of adult stem cells found in muscle tissue but also provides a protocol for their successful use in restoring function to damaged organs.

Circulation

Scientists in Dresden, Germany, have succeeded in coaxing adult stem cells to differentiate into endothelial tissue.[29] This news could lead to new treatments for those suffering with poor circulation and tissue damage. The researchers took adult stem cells from human bone marrow and cultured them *in vitro* adding endothelial growth factors. This triggered the adult stem cells to develop into types of cells that form blood capillaries. This research is especially important for those studying circulation. Endothelial tissue is crucial in the formation of new blood capillaries. Using adult stem cells to enhance the flow of blood to specific parts of the body could lead to new treatments for poor circulation and internal tissue damage. Scientists at the National Cancer Institute, National Institutes of Health in Bethesda, Maryland, have also published evidence that umbilical cord blood harbors stem cells with the same exciting potential. Using a similar technique, the

American team harvested stem cells from cord blood and coaxed them to develop into endothelial tissue. Despite the use of alternative sources of adult stem cells, the results appear to demonstrate the versatility of adult stem cells to treat a myriad of medical disorders.

Also with regard to circulatory deficiencies, the plasticity (ability to become any cell of the body's tissue) was demonstrated again by Andrew E. Wurmser et al, in *Nature*, July 15, 2004; neural stem cells were developed into cells that line blood capillaries. The team took adult stem cells from mice and co-cultured them with human vascular tissue. Six percent of the cells originating from the mice developed into vascular tissue. The result is said to be particularly surprising as nerve and vascular cells are only distantly related. Researchers suggest that this work may open the door to alternative treatment strategies for those with circulatory deficiencies.

General Body Tissue

Scientists at a national meeting at the University of Virginia Health System announced findings from a significant number of studies that adult stem cells from adipose tissue (fat) could eventually be used to treat other damaged or injured tissues. In total, forty-seven research abstracts were presented from both academia and the private sector suggesting that adipose-derived stem cells can be used to repair or regenerate new blood vessels, cardiac muscle, nerves, bones, and other tissue, potentially helping heart attack victims, patients with brain and spinal cord injuries, and people with osteoporosis. The work presented reflected the belief of a growing number of researchers that adipose (fat) would be a practical and appealing source of stem cells for regenerative therapies of the future.[18]

Brain Tumors

Adult stem cells found in the skin may soon provide a novel treatment for patients with brain tumors. Meeting at the fifty-fourth annual conference of the Congress of Neurological Surgeons, researchers from Italy announced their success in isolating adult stem cells harvested from skin samples. The next step will be to implant the adult stem cells into the brains of patients suffering from tumors, with the hope that the treatment will stimulate growth of new blood vessels and suppress tumor development. The work follows promising studies carried out on mice, which showed that implanted adult stem cells enhanced blood supply and led to decreased tumor growth. Mice treated in this manner survived an average of 50% longer than their untreated counterparts.[19]

A significant piece of new research has come from Dr. Catherine Verfaillie's laboratory in the Stem Cell Institute at the University of Minnesota. Dr. Verfaillie reported the isolation of human bone marrow cells capable of changing into multiple cell types. Verfaillie's adult stem cells, termed multiple adult progenitor cells, MPACS, are the first to be isolated and grown in culture and be as potentially versatile as embryonic stem cells. These MPACS, isolated from donated bone marrow samples, grow indefinitely in culture and can be coaxed to form muscle, bone, liver, and various neurons and brain cells. These cells have not been directly compared with embryonic stem cells, but the early indication of this is promising.[23]

Uterus

At the Monash Institute of Medical Research in Australia, senior scientist Dr. Caroline Gargett's discovery of adult stem cells in the uterus that can be grown into bone, muscle, fat, and cartilage has been hailed as a major medical development by international reproduction experts. Dr. Gargett explained how two types of adult stem cells have been extracted from endometrial tissue in the uterus. "While adult stem cells have been found in other parts of the body, no one has ever identified them in the uterus before. Not only will this assist with understanding how several diseases of the uterus develop, but also could also contribute to general studies into adult stem cells.[20]

"The discovery of mesenchymal stem cells in the uterus is particularly significant as it is this type of stem cell that bone, muscle, fat and cartilage are formed," she said. "We can now grow these tissues in the lab and are investigating avenues to apply the technology."

The initial focus of the team at the Monash Institute of Medical Research was in using these stem cells to aid in the repair of pelvic floor prolapse. "If we could offer women a bioengineered ligament that is made from their own stem cells, the long term quality of life for thousands of women who suffer this problem could be greatly enhanced," she said.

One in ten women require treatment for uterine prolapse, usually in their fifties and older, although it can happen to younger women. The pelvic floor is weakened during pregnancy and childbirth, and as a woman ages, the strength of these muscles can deteriorate further. "At present, we use surgery to repair prolapsed uterus, which is a form of a hernia," she said. "However, in almost 30% of women the prolapse can reoccur. In order to reduce this chance of reoccurrence, a reinforcement material, often a synthetic mesh, is applied. While this technique can be successful, complications also frequently arise due to erosion or rejection of foreign matter. A firm natural tissue made of adult stem cells certainly would be advantageous."

Liver

Adult stem cells are also rebuilding livers. Until now, the only hope for persons with irreversible liver failure from such diseases as cirrhosis, which kills about 27,000 Americans yearly, was transplantation. This requires permanent use of immuno-suppressive drugs that can lead to opportunistic infections and cancer. Most importantly, it requires a new liver. About a thousand Americans are now on a waiting list for a liver transplant, and many will die while on the list. But London's Imperial College reports in *The New Scientist* that they have repaired patients' own damaged livers by using bone marrow adult stem cells painlessly collected from their own blood. Five were injected with a drug that stimulated their bone marrow to produce extra stem cells that were injected into a blood vessel leading directly to the liver.[21]

It worked. Both the liver function and overall health of three of the five treated patients improved significantly within only two months of treatment. The two patients whose health did not improve were left no worse off.

The researchers said the bone marrow stem cells appeared to simply home in on damaged portions of the liver and affect repairs, just as adult stem cells have been shown to do with other organs thought to be unable to repair themselves such as hearts and brains.

In another study, the liver was rejuvenated using adult stem cells. Scientists at Johns Hopkins University, Baltimore, Maryland, observed improvements in mice suffering with liver damage within days of giving the animals adult stem cell implants. Further examination revealed that the stem cells, which had been harvested from bone marrow, had responded to chemical cues within the damaged tissue and developed replacement liver cells. The breakthrough, published in *Nature Cell Biology*, June 2005, further broadens the horizons of adult stem cell technology and gives hope to the thousands of people suffering with liver damage.

Spine

ABV News Online, on November 28, 2004, reported that a South Korean woman paralyzed for twenty years is walking again after scientists repaired her damaged spine using stem cells derived from umbilical cord blood.

Hwang Mi-Soon, thirty-seven, had been bedridden since damaging her back in an accident two decades ago. South Korean researchers recently went public for the first time with the results of their stem cell therapy. Ms. Hwang walked into their press conference with the help of a walking frame. The researchers say it is the world's first published case in which a patient with spinal cord injuries has been successfully treated with stem cells from umbilical cord blood. The research needs verification from international experts, but the case could signal a leap forward in the treatment of spinal cord injuries. Use of stem cells from cord blood also points out a way to side step the ethical debate over the controversial use of embryos in embryonic stem cell research.

Scientists in Russia have announced success in using adult stem cells to treat six patients suffering from spinal injury.[23] The results mirror those of Korean scientists, who performed similar experiments using stem cells from umbilical cord blood, and the work of a Portuguese researcher, Dr. Carlos Lima. The six patients involved were all bedridden and thought they would never regain use of their legs. However, following stem cell transplant surgery, each of them is now re-learning how to walk.

It was Andrei Bryukhovetsky, director of the Neurology Clinic, who suggested performing the operations involving transplantation of stem cells to the spinal cord. Extracted neural stem cells are grown in tissue culture then injected into the damaged area of the spine, restoring the nerve function of the vertebrae one by one.

Eye

A team of collaborators from Canada and Switzerland announced the successful isolation and changing of adult stem cells taken from the human eye.[24] When cultured *in vitro*, out of the body, the stem cells divided and took the identities of retinal cells. What's more, when implanted into the eyes of mice, the cells were observed to migrate appropriately, differentiate, and then integrate at the right developmental times.

The researchers' findings could lead to a cure for some types of blindness such as retinitis pigmentosa, an inheritable disease of the retina characterized by night blindness and gradual loss of peripheral vision, caused primarily by the degeneration of the photoreceptors of the retina. The findings also have implications for treating macular degeneration, a disease affecting central, sharp vision through damage to the macula, which is part of the retina.

Blood Disorders, Diabetes

The treatment advantage in using a patient's own adult stem cells to generate replacement tissues required to reverse serious diseases is that the implanted tissues are immunologically identical to the patient, thus avoiding any risk of donor rejection. A problem is that any genetic flaw is simply transferred by the stem cells to the implanted tissue, and the disease is perpetuated. Pioneering research carried out at the University of South Florida has led to a solution.[25] Scientists investigating alpha 1-antitrysin deficiency, a genetic liver disorder, took adult stem cells from mice suffering the condition. They then enhanced the cells *ex vivo,* out of the body, by incorporating the human gene required to produce the functional version of the enzyme. The cells were then implanted back into the diseased mice. The enhanced cells incorporated themselves into the mice's livers, repairing damaged tissue and restoring normal physiological functioning. The research team observed long-term engraftment of the implanted tissue and saw no evidence of transgene instability. This protocol represents a way of delivering genes to diseased organs. The implanted material is immunologically identical because it derives from the patient, yet it carries a functional copy of the required gene. This procedure could be used to treat human patients suffering from conditions ranging from blood disorders to diabetes.

Stroke

Also at the University of South Florida, researchers in the College of Medicine have used umbilical cord stem cells to treat patients with stroke damage.[26] The team treated rats twenty-four hours after they had suffered from a stroke. By monitoring the animals' progress, scientists showed that the treatment led to substantially improved behavioral abilities in rats receiving the treatment, when compared with rats left untreated. Further analysis of the animals' physiology revealed that the stem cells present in the umbilical cord blood had migrated to the brain and became localized only in tissues damaged by the stroke. The findings add to previous evidence gained using umbilical cord blood and further support those working toward clinical trials using the blood to treat stroke patients.

As reported by the *Reno Gazette*[27], a former judge who suffered a stroke is being given the chance for a normal life by a team of Ukrainian scientists. Judge Mills Lane, who suffered the stroke while at home in Reno, Nevada, was left with impaired speech and unable to use his right arm. Now, a team based in Kharkov, Ukraine is using Lane's own adult stem cells in an attempt to improve his coordination. The scientists took stem cells from Lane's bone marrow and injected them into his central nervous system. It is hoped that the stem cells will help repair nerve tissue damaged by the stroke. Mills, who first received stem cell therapy in August 2003, has already shown signs of improvement and has returned to Kharkov to complete the treatment. The results are seen as especially exciting as Mill's body has completely accepted the injected tissue. This is most likely because he himself is the tissue donor as well as the recipient, so the injected stem cells are a perfect immunological match.

Arthritis

S.V. Pavietic reports in *Arthritis and Rheumatism 50* that his research team has successfully used adult stem cells to treat a fifty-two-year-old woman suffering from rheumatoid arthritis. The patient, who received adult stem cells from her sister, showed a total reversal from rheumatoid arthritis. Her rheumatoid nodules completely disappeared after only twenty-one months of treatment; she is disease-free and is not taking any drugs to suppress her immune system. Doctors in charge of the operation now intend to use the procedure to treat other patients, saying they are satisfied the transplant may be performed safely, without the development of graft-versus-host disease or serious infection, and results in marked resolution of rheumatoid arthritis.

Recently, bone marrow stem cells have been shown to be more versatile than ever. For some time, bone marrow stem cells have been known to change into heart muscle, kidney tissue, lung tissue, liver, and skin as well as the full range of blood and marrow cells. But, until recently, critics of adult stem technology have argued that cells observed in animal experiments are the products of cell fusion, not genuine new cell development. Their claim is that rather than developing from stem cells into other tissue types, implanted stem cells simply merge with pre-existing cells and adopt their characteristics. Findings by a team at Yale University in Connecticut refute this model. The team tested the fusion hypothesis by implanting bone marrow stem cells into female mice. Using a number of analytical techniques, the researchers found that stem cells developed into a range of different tissues. The presence of genetic markers and the cell's morphology, or source, ruled out the possibility of fusion. With this explanation eliminated, the team concludes that bone marrow stem cells are capable of developing directly into a wide range of different tissues. The finding supports those who identify adult stem cells as the most likely source of treatment of incurable diseases.

Chapter Nine

Find Your Own Stem Cell Clinical Trial

Before trying to find an appropriate stem cell clinical trial for yourself or a loved one, make certain that you are already receiving the best possible conventional treatment. If you have not already done so, it is appropriate to get a second opinion of your diagnosis and treatment. Your doctor will usually not be at all upset if you request a second opinion and will be happy to refer you to another doctor or to a medical center at a teaching hospital.

An alternative to getting a referral from your doctor is to seek an opinion from the best doctor specializing in your particular disease. There are books containing information on outstanding doctors at your library, at a local bookstore, or on the Internet. A quick survey of Amazon.com yielded several such books. Here is an example:

America's Top Doctors: Choosing the Best in Healthcare, 4th Edition, by John J. Connolly.

Use any of the many search engines on the Internet to find clinical trials of stem cell therapies. There are about one hundred search engines available on the Internet, but you can find almost everything by going to just a few. Over 80% of searches are made on Google, Yahoo, AltaVista, and AOL. Google is probably the best place to start. Google has even become a verb in the dictionary. It is now correct to say that you are "googling" stem cell clinical trial information.

Unfortunately, when you perform your search on Google or any of the other search engines, the results may not be dependable information. The search will bring you everything: all the trash, commercial sites, outdated pages, and everything else you can imagine. Nothing will have been checked for credibility or accuracy. You may have to sort through thousands of Websites to find a few that will be useful. A new feature is the prominence of paid Websites at the top and sides of Website listings. Commercial companies pay for each "hit" or time that their Websites are accessed. Some of the information you find when going to one of these commercial sites may prove to be useful, but keep in mind that the purpose of the site is to sell a service or product, not to merely provide you with information.

Avoid deceptive ads and information. Beware of "miracle" treatments and cures. They can cost you money and your health, especially if you delay or refuse proper treatment. Here are a few indications that a product is truly too good to be true:

- Phrases such as "scientific breakthrough," "miraculous cure," "exclusive product," "secret formula," or "ancient ingredients."
- Claims that the product treats a wide range of ailments.
- Use of impressive-sounding medical terms. These often cover up a lack of good science behind the product.
- Anecdotal case histories from consumers claiming "amazing" results.
- Claims that the product is available from only one source, or for a limited period of time.

During the course of your research, you will find stem cell clinical trials that are being conducted overseas. It is wise to carefully question the efficacy of these trials or treatments and to investigate the source and quality of qualified medical supervision. Although overseas stem cell treatments and clinical trials may have been suspect in the past, you will now find research centers at major hospitals around the world that are conducting legitimate clinical trials of stem cell treatment. Some may offer treatments using embryonic stem cells that are not available in the United States.

Note that most American stem cell clinical trials are free except for travel and living expenses outside the hospital. Overseas treatments are likely to be expensive, and they are not covered by most insurance plans.

It is a good idea to narrow your search as much as possible. For example, if you conduct a search for "leukemia" using Google, you will find approximately forty million Webpages containing that word. Obviously, this number is unmanageable and you must reduce your search by being more specific. For example, "leukemia therapy" will reduce the number of articles to about 86,000 Webpages containing that combination of words. Note the addition of quotation marks at the beginning and ending of the words. This limits the number of pages to those containing both words in sequence.

You may want to be even more specific. If you search under leukemia + stem cell therapy, you will find over four million Webpages containing those words. Although this number is huge, you should open and start scanning as many Webpages as you can. Your work is compounded further by the need to follow every link at every Webpage. A link is a spot that you click on with your mouse that will lead you to related Websites. It is important that you follow these links, because the Webpage at the end of any link may have the exact information you are seeking.

A great deal of work is involved, but you need to be as diligent as possible to avoid overlooking information concerning a stem cell clinical trial at a recognized medical center.

Several credible medical Websites provide accurate information concerning the availability of stem cell clinical trials. Trials for any number of serious diseases are listed. These Websites are published by several U.S. government agencies and by private agencies such as the American Cancer Association. Give these sites your highest priority. However, from my own experience, I have learned that you should not overlook the value of a general search using one of the search engines. I found that not all available clinical trials are listed on the government and private agency Websites. A legitimate stem cell clinical trial may be mentioned in a newspaper, magazine, or medical journal and not be found in the listings made available by the National Institutes of Health and other authoritative Websites. That having been said, the following Websites should be included in your search:

- Agency for Healthcare Research and Quality, www.ahrq.gov/consumer

The Agency for Healthcare Research and Quality (AHRQ) is the lead federal agency charged with improving the quality, safety, efficiency, and effectiveness of healthcare for all Americans. As one of twelve agencies within the Department of Health and Human Services, AHRQ supports health services research that will improve the quality of healthcare and promote evidence-based decision-making. This agency primarily provides information for healthcare professionals but also has information for consumers.

- CenterWatch Clinical Trials Listing Service, www.centerwatch.com

This is an information source for the clinical trials industry. You can use this site to find a wealth of information on more than 41,000 active industry and government-sponsored clinical trials, as well as new drug therapies in research and those recently approved by the FDA.

- You can also get information on clinical trials at the government's NIH (National Institutes of Health). This site provides an introduction to clinical trials, information on participation in clinical trials, a glossary of clinical trial terms, and links to information on clinical trials. You can also browse by condition, sponsor, and recruitment status. National Institutes of Health, www.clinicaltrials.gov

- Cochrane Collaboration, www.cochrane.org

This non-profit organization is dedicated to making accurate information about the effects of healthcare readily available worldwide. A listing of clinical trials can be obtained by using the Website's search engine.

- Trials Central, www.trialscentral.org

This non-profit agency provides a service for providing clinical trial information after entry information concerning the patient's illness.

- PubMed, www.ncbi.nlm.nih.gov

PubMed is published online by the National Library of Medicine. You can search on this site for information for any particular disease or condition; however, the papers are highly technical, and you may need your doctor to explain them to you.

- Medscape, www.medscape.com.

Medscape, produced by WebMD, emails information to you on new therapies as it becomes available. You must register for it at Medline Plus. Medline Plus will also send you updated medical information as it becomes available. You can register for it at www.nlm.nih.gov/medlineplusnews. This is also a service of the government's National Institutes of Health.

Your literature research will reveal stem cell clinical trials for almost all serious diseases. I urge you to follow up and become included in an early-on advanced treatment that might ease your suffering or save your life.

After locating stem cell clinical trials that may seem to be beneficial for your illness, your next step will be to contact the trial coordinator. You can usually do this by letter, phone, fax, or email. I would suggest that you use the phone so that you can immediately get additional information about the trial and be pre-screened for inclusion in the trial. This will save a great deal of time, which is extremely important when dealing with a serious illness.

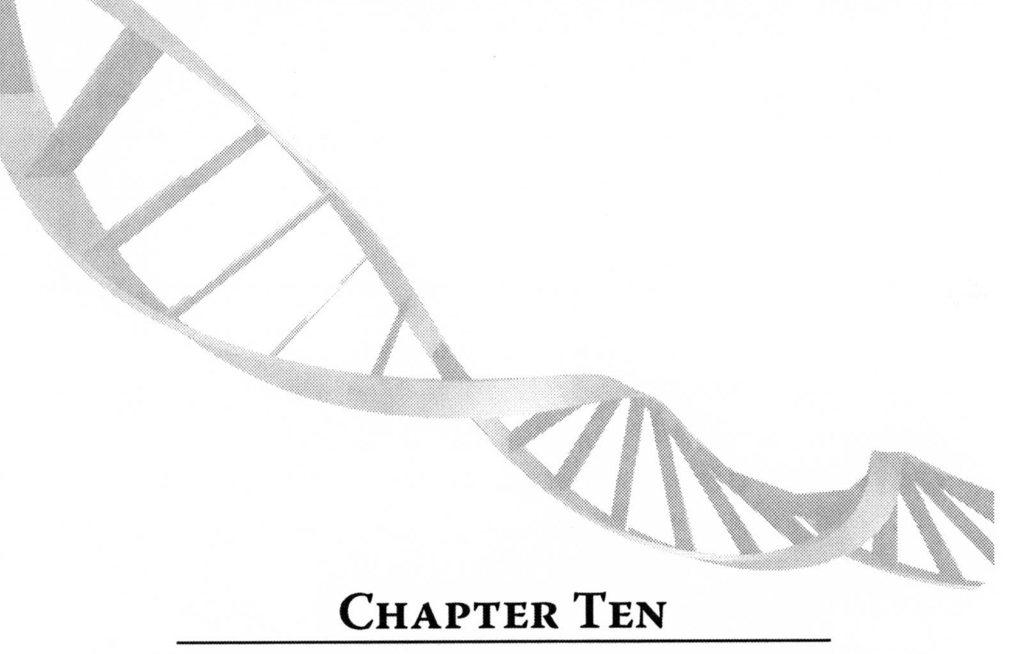

Chapter Ten

New Stem Cell Therapies

By the time you read this book, the number of stem cell studies and clinical trials will have undergone an explosive growth. Initial testing in the laboratory for a multitude of diseases may have advanced to Phase I clinical trials using human subjects. Phase I trials may have advanced to Phase II and Phase III. All of these trials will require many volunteer subjects. In addition, the requirements for admission into later clinical trial phases will become less and less stringent because scientists will evaluate patients with a broader scope of a disease's progression.

According to the National Institutes of Health, stem cell therapy is presently in clinical trials for a multitude of diseases including: diabetes, ovarian cancer, and childhood non-Hodgkin's lymphoma, leukemia, mastocytosis, kidney cancer, colon/rectal cancer, esophageal/gastric cancer, urinary cancer, sickle cell disease, lung cancer, breast cancer, aplastic anemia, brain tumor, HIV, peripheral vascular disease, and many other conditions.

There is hope that stem cell therapy can help sufferers of many additional diseases. It is up to you to discover what treatments are available and where they can be obtained. Whatever you do, don't ever give up hope.

The Latest Research on Stem Cells and Gene Therapy

Gene therapy refers to a method of treating diseases caused by defective or absent genes. It can also describe the addition of new genes that have been scientifically designed to fight diseases. Such therapies are closely related to stem cells and stem cell therapies. Scientists at Johns Hopkins published an article in the January 2002 issue of *Blood* that described a method of using stem cells as a vehicle for gene therapy. The researchers worked with mice bred to have poor immunity. They injected those mice with blood stem cells that carried a specific gene that is activated only inside a specific mature immune cell, called an antigen-presenting cell. They were able to monitor the fact that the stem cells ultimately developed into the antigen-presenting cell by tagging the gene in question with a fluorescent molecule and measuring the amount of fluorescence in the offspring cells. The scientists demonstrated that planting a gene in a stem cell and then transplanting the stem cell could indeed result in that gene being found in the mature cell population that needed it. Among the potential uses for this type of gene therapy would be the insertion of genes into stem cells that could boost the immune system and perhaps allow for more effective anti-cancer vaccines. It could conceivably modify the immune system such that transplants were easier to accomplish without organ rejection. This technique might also be used to overcome the potential rejection of stem cell transplants themselves.

Recently, a group from the Goldyne Savad Institute of Gene Therapy in Jerusalem, Israel, showed that a lentviral vector (an agent that transfers genetic material from one cell to another) could be used to modify the gene expression of human embryonic stem cells in the laboratory. The modified gene expression continued as the stem cells proliferated and differentiated, demonstrating the technical feasibility of using stem cells for gene therapy.

The following examples will give you an idea of the large number of therapies for various diseases currently in early stem cell clinical trials. This information will become dated as time passes, and you need to do your own search for current information.

Cancer

Melanoma is a potentially lethal form of skin cancer for which there has been no effective treatment for later stages except surgery. Presently, clinical trials using stem cells as a part of a therapy for treatment of advanced melanoma are being conducted in twenty-two states around the country. Many of these trials are still recruiting patients.

Northwestern University researchers have demonstrated how the microenvironment of two human embryonic stem cell (hESC) lines (federally approved) induced metastatic (moving from one organ to another) melanoma cells to revert to a normal, skin-like type with the ability to form other colonies of hESCs. The researchers also showed that these melanoma cells were less invasive following culture on the microenvironments of hESCs.

"Our observations highlight the potential utility of isolating factors within the human embryonic stem cells' microenvironment responsible for influencing tumor cell fate and reversing cancerous properties of metastatic tumor cells such as melanoma," said Mary J. Hendrix, in whose laboratories at Children's Memorial Research Center the experiments were conducted.

Heart

There is a very exciting new approach in the use of adult stem cells from bone marrow for repair of heart damage caused by heart attack. As background, in my own experience, a type of stem cell called CD34-positive was harvested from my bone marrow and injected into my heart muscle. The purpose of the CD34-positive type of cell is to stimulate an increase of blood flow to oxygen-starved heart muscle with the purpose of relieving heart pain. Additionally, there are some indications that muscle growth is stimulated. CD34-positive stem cells may even emit chemical signals, which induce healing.

An exciting new discovery in stem cell therapy for the heart is the use of adult mesenchymal stem cells, which are also harvested from bone marrow. When injected into the heart during the course of open-heart bypass surgery, the mesenchymal stem cells immediately begin to stimulate growth of heart muscle to replace tissue damaged during a heart attack. Dr. Silviu Itescu at the University of Pittsburgh says, "Maybe after therapy by stem cells they won't need heart transplants." Dr. Itescu is cofounder of Angioblast Systems, a biotech company that focuses on mesenchymal precursor stem cells to repair cartilage, bone, and muscle.

The first direct evidence that muscle cells transplanted from within a heart patient's body could help heal their damaged heart muscle is being reported by a team from the University of Michigan Health System, Massachusetts General Hospital, and Diacrin, Inc.

The results come from three patients who had cells from their thigh muscles injected into their heart muscle while they awaited a heart transplant; they then allowed their old, damaged hearts to be examined for signs of cell growth after they got a new heart. The detailed examinations, the first of their kind, showed that the injected cells not only survived in their new environment but also began to form muscle fibers. The areas where the cells were injected also had a small increase in the formulation of small blood vessels. None of the patients experienced immune reactions to the cell transplants.

"These results give us the first indication that muscle cell transplants, even from an entirely different kind of muscle, could one day be used to help repair damaged, failing hearts without danger of rejection," says University of Michigan cardiac surgeon Francis Pagani, MD, PhD. "We have further to go, but we are very encouraged." The detailed findings come from one arm of a two-part Phase I study sponsored by Diacrin, designed to see if transplanted skeletal muscle cells might be a feasible option for repairing hearts damaged by heart attack and other diseases.

The study rests on the premise that certain kinds of cells can be expanded in culture and maintain their functional characteristics. Though similar to the concept of stem cells that become any kind of cell in the body, the study uses "satellite cells," which occur naturally in muscle and help repair damage by dividing and moving to injured areas. The arm of the study conducted at the University of Michigan Health Center and Temple University involves patients awaiting a heart transplant to replace their scarred, failing hearts. The patients were also scheduled to receive an implanted heart-assisting device called an LVAD to help them survive until a new heart becomes available. Because their old heart can be removed for tests after the transplant, detailed analysis of the injected cells is possible.

The other arm of the study, led by the Arizona Heart Institute, looked at how well patients tolerate different doses of transplanted skeletal muscle cells injected during heart bypass surgery. It assessed the safety of the cell injection and found indirect evidence of scar tissue regeneration. However, they could not examine the heart directly.

In both arms of the study, a sample of cells was removed from the quadriceps muscle and treated with enzymes to isolate the satellite cells. They were grown in a laboratory under carefully controlled conditions to give the original handful of cells time to divide and produce 300 million cells.

The surgeons injected the cells into the wall of the heart's pumping chamber during an open-heart surgical procedure, either the LVAD implantation surgery or a coronary artery bypass graft (CABG) operation.

In LVAD patients, the new cells were placed in cardiac muscle tissue that had been severely scarred and hardened to the point that it could no longer contract sufficiently to help pump blood. The LVAD helps boost the patient's pumping power by feeding blood into a battery-powered metal pump and out through a tube connected to the main artery. University of Michigan Health Center teams have implanted more than one hundred patients with the type of LVAD called a Heartmate in the past six years.

When the patient got a heart transplant, the old heart was removed and sent for a series of histological tests at Diacrin headquarters in Charlestown, Massachusetts. The new results come from an analysis of two University of Michigan Health

Center patients and one Temple patient. Two other University of Michigan LVAD patients have received cell injections but are still awaiting heart transplants.

"The results show direct evidence of skeletal muscle survival and differentiation into mature muscle fibers, measuring using antibodies that specifically target skeletal muscle cells," Dr. Pagani says. "Because cardiac muscle and skeletal muscle are two distinct types of tissue, the antibody tests show conclusively that the transplanted skeletal muscle satellite cells survived."

The transplanted cells also appear to have begun forming vascular muscle cells, which make up the walls of blood vessels. In areas where cells have been injected, there was a significant increase in small blood vessels compared with areas that had not received injections.

In addition to the encouraging finding that the injected cells "grafted" into their new environment, the results show on a molecular level that the heart muscle did not reject the newcomers. No evidence of an immune reaction or lymphocytes was seen in either grafted or non-grafted areas, using a test specific for T-cells that usually respond to "invasions" of foreign cells.

"Because the skeletal muscle cells are from the patient's own body, we don't expect the kind of immune reaction and rejection that we often see in transplants of whole hearts from donors," says Dr. Pagani. "If further study bears this finding out, we may have a new option for repairing hearts without putting the patient at risk of dying from rejection, or needing life-long anti-rejection medications."

Dr. Pagani stresses that these early results, while encouraging, are merely the first steps in evaluating skeletal muscle cell transplants. Combined with the results from the safety wing of the study, the initial results may help the researchers determine how to proceed toward evaluations of whether cell transplants can actually help patients' hearts.

"The promise of this line of research is immense, but we must be careful not to overstate what we have found thus far," says Pagani.

"Only through further research and the cooperation of more LVAFD and bypass patients can we see whether we can get a clinically significant effect."

Stroke

At AIIMS, a research hospital in India, stroke victims are being treated with stem cells. For example, a seventy-year-old patient who suffers neurological disorders that restrict his movements received stem cell therapy with great success.

Cancer

Texas researchers say they have perfected a method to deliver cancer treatment directly into tumors, bypassing healthy tissue.

The study was done on mice, but human trials could soon begin, said Dr. Michael Andreef at the University of Texas Anderson Cancer Center, one author of the study reported in the *Journal of the National Cancer Institute.* The research team used the benefits of a known anticancer therapy called interferon beta that can kill cancer cells. In practice, the therapy has proven problematic. It causes toxic side effects, and its benefits disappear within minutes of patients getting their shots. The research team worked around those problems by manipulating a certain type of stem cell to encode the interferon beta gene. The stem cells then move around like guided missiles, targeting tumor cells and producing high concentrations of therapeutic protein within the tumor cells.

Besides taming toxic side effects, the cancer treatment stuck around the tumor longer. Mice with human breast cancer treated with engineered human stem cells survived for sixty days. Mice treated with the interferon alone lived for forty-one days. Untreated mice survived for thirty-seven days. Meanwhile, mice with melanoma treated with the stem cells survived seventy-three and a half days compared with thirty days for untreated mice.

A clinical test is being planned in which human patients would be infused with the stem-cell-delivered treatment four times a week. The targeted delivery of anticancer therapy to tumors builds on what researchers already know about how wounds heal. The specialized stem cells, known as mesenchymal stem cells, come from bone marrow and help maintain healthy connective tissues. When new tissue is needed to heal wounds or form scars, those special cells swell in number.

Even though they are tumors, the malignant cells act like never-healing wounds. The research team did not see engineered stem cells drift into healthy organs like the lungs, liver, spleen, kidneys, or muscles. The fact that the stem cells are driven by a duty to help means that a wound elsewhere in the body could distract some cells from reaching tumors. Any wound that is active, that requires repair, would be a target. That means that doctors would need to screen patients carefully to ensure the therapy is not attempted on people who had recently undergone surgery.

Bladder

According to the British journal *Lancet*, researchers reported that they have grown complete urinary bladders in the laboratory and transplanted them into seven children with spina bifida, improving their health and cultivating the first replacements for failing organs of people. The "neo-bladders," each one grown in a small laboratory container from a pinch of the patient's own stem cells, have been working in seven young patients for an average of almost four years.

If ongoing studies continue apace, the researchers hope someday to offer patients more than a dozen other home-grown organs, including blood vessel complexes, partial kidneys, and perhaps hearts.

Because the replacement bladders were made from the patient's own stem cells, they did not stimulate the immune system reaction allowing them to completely avoid the risks of rejection.

Parkinson's Disease

Certain neurological diseases involve the destruction of nerve cells. Such diseases seem to be prime candidates for treatment by stem cell transplantation. If multipotent (able to become many types of tissue) stem cells could be successfully encouraged to migrate to areas of damaged or dead nerve cells and replace them with new and healthy cells, those neurological diseases could theoretically be improved or perhaps cured. One area in which much research has been focused is Parkinson's disease. In Parkinson's disease, the degeneration of a group of nerve cells that produces the chemical transmitter dopamine leads to a movement disorder featuring tremors.

Transplantation of fetal nerve (brain) tissue directly into the brains of Parkinson's disease patients had been done in a few centers, with varying results. Because the tissue is typically derived from products of legal abortions, the procedure has been controversial. Scientists at a number of institutions are performing a variety of animal studies with stem cells from different sources, but most of the human procedures performed have used fetal tissues.

Experiments conducted at the National Institutes of Health have demonstrated that embryonic rat nerve cells grown in culture and transplanted into rats with a Parkinson's-like disease will then differentiate into healthy, mature brain cells. The rats that underwent this procedure experienced some neurological recovery. Researchers in Britain have isolated human embryonic nerve cells, grown them in cultures in the presence of substances known as growth factors, and then transplanted those cultured cells into the brains of rats with Parkinson's-like diseases. At autopsies of the rats twenty weeks later, those human cells were located far from the original transplant sites. In addition, they were found to have differentiated into several types of adult rat brain cells.

The successful transplantation of human fetal cells into animals gives hope that fetal nerve cells from other species could be transplanted into humans, with some recovery of neurological function. Using fetal tissue from other species would eliminate the controversy surrounding the use of human fetal tissue. Researchers at Harvard Medical School have transplanted fetal pig neural tissue into the brain of a human with Parkinson's disease. After the death of that patient, apparently of causes unrelated to the transplant, an autopsy was performed. It showed that the pig cells had matured into cells that would produce dopamine, the missing chemical transmitter in Parkinson's disease. The researchers suggest that further refinement of that technique may one day offer hope for humans.

Work continues in this field. Research published in January 2002 in the *Proceedings of the National Academy of Science* from Harvard Medical School detailed more animal experiments in this area. The scientists transplanted undifferentiated mouse stem cells into the brains of rats with a Parkinson's-like syndrome. The cells differentiated into fully mature neurons that produced dopamine, the missing chemical messenger of Parkinson's disease. The animals showed functional improvement as well. While these results are encouraging, more work needs to be done before this technique becomes accepted procedure for humans with Parkinson's disease.

In later studies, stem cell implant surgery is offering hope for some people with Parkinson's disease. In a study of this new experimental procedure, patients under age sixty regained some movement after having stem cells implanted. The report appears in the April 2006 issue of *Archives of Neurology.*

"We should be very encouraged...by the margins of improvement," writes Roger N. Rosenberg, MD, the journal's editor, in an accompanying editorial. While more work needs to be done, these findings "...provide an important beginning..." in developing highly specific, gene-based therapies in treating Parkinson's disease. "The golden age of Neurology is just beginning!"

Parkinson's disease is a degenerative disease of the brain—one that has no known cause or cure, writes lead researcher Paul H. Gordon, MD, a neurologist with Columbia-Presbyterian Medical Center in New York. Parkinson's disease impairs and slows voluntary movement and contributes to disability.

With new imaging and improved surgical techniques, there has been renewed interest in surgically treating Parkinson's disease. Attempts were made to slow the progression of Parkinson's and even reverse its course with implantation of stem cells, notes Gordon. The study used stem cells that were coaxed to produce dopamine, the brain chemical that stimulates brain nerves. People with Parkinson's disease have a deficiency in this chemical, which leads to the slowed movement characteristic of this disease.

In the study, twenty patients with advanced Parkinson's disease were randomly chosen to have either surgery in which the brain cells were implanted in their brains or sham surgery in which nothing was implanted. After one year, patients who got stem cell implantation showed "significant and lasting" improvement in hand and foot movement compared with the deterioration seen in patients who received the sham procedure—especially those patients under age sixty, he writes.

Older patients who did not have stem cell implantation surgery had significant deterioration during the same period, reports Gordon. That group was sixty years and older, and the deterioration may have been related to their age.

Spinal Cord Injury

Myelin is a protective sheath that surrounds nerve cells. When myelin is damaged, as may occur in many spinal cord injuries, the underlying nerves are damaged, often apparently beyond repair. Regeneration of myelin offers a theoretical method of reversing spinal cord injuries.

A study, published in the *Proceedings of the National Academy of Sciences*, was conducted at the Washington University School of Medicine in St. Louis to look at this possibility. Chemicals were applied to the spinal cords of a population of laboratory rats, which dissolved their myelin. Three days later, rat embryonic cells were transplanted into the rats' spinal columns. When the rats were sacrificed and autopsied, mature myelin-producing cells were found at the site of the transplants. When these same embryonic cells were transplanted into rats that were genetically deficient in myelin production, they too were found to have mature versions of the embryonic transplanted stem cells, and those matured stem cells were producing myelin. Further work in the same laboratory was done on rats with induced spinal cord injuries. Nine days after injury, those rats were treated with embryonic stem cell transplants. After two to five weeks, the rats demonstrated improvements in weight bearing and coordination, and at autopsy, they were found to have adult versions of the transplanted fetal cells. If these types of procedures are someday available to humans, their benefits to spinal cord injury patients could be enormous.

Research that is more recent was presented at the April 2001 meetings of the American Association of Neurological Surgeons. Fifteen female mice with spinal cord injuries were studied. Eight were treated with transplants to their spinal cords of stem cells of neural origin, and the other seven were used as controls. At day seven after transplant, the mice were assessed for neurological function. The mice that had received transplants showed significantly greater functional recovery than the control mice.

Recently, researchers found the white matter of the adult human brain, which is made up of myelinated nerve fibers, seems to harbor stem cells that can become glia, a type of white matter, or neurons, which are the grey matter. The cells successfully differentiated, both in vitro and after transplantation to fetal rat brains. Published in *Nature Medicine*, these results could potentially be used to repair both white and grey matter.

A leading British neuroscientist believes he will soon be able to treat patients with spinal cord injuries using stem cells. Professor Geoffrey Raisman at the Institute of Neurology at the College of London has pioneered a technique that involves transplanting adult stem cells from the lining of the nose into an area of injury of the spinal cord. His team has carried out tests on rats with injuries to a specific small tract in the spinal cord. His team has been able to restore use of the paw for retrieving food and climbing and use of the diaphragm. They are now looking at human adult stem cells and expect to be seeing human patients.

The traditional scientific view has been that after damage to the brain or spinal cord, the body had no ability to regenerate the connections. Professor Raisman now believes that after injury, new connections form automatically. He believes the problem is that nerve fibers, which have been cut, do not regenerate; new connections are additional ones formed by existing nerve fibers in the area. And they do not regenerate the circuitry. They simply restore the gaps. The idea is to reopen the nerve pathways by reorganizing the cells of the scar tissue so the nerve fibers will grow back. A circuit is then reconnected to restore function or relearn, even if it is not the original circuit.

The aim is to repair spinal cord injury in humans by transplanting stem cells from the nasal lining called olfactory ensheathing cells (OECs) into areas of injury. These cells are chosen because the nasal lining is the only area of the body where nerve fibers are known to be able to grow throughout adult life.

Professor Raisman is proposing that the first human trials are carried out on patients with spinal cord injury known as brachial plexis avulsion, where nerves to the arm are pulled out of the spinal cord, frequently as the result of car accidents. He also believes that, in time, the procedure can be applied to more severe spinal cord damage and to other injuries such as stroke, blindness, and deafness.

Multiple sclerosis is a disease in which demyelination, or loss of myelin, is a cardinal feature and the cause of neurological defects. Researchers are asking if stem cell therapy that induces remyelination might slow or reverse the neurological problems of multiple sclerosis. Researchers at Emory University in Atlanta published results of such a study in *Nature Magazine.* The researchers had transplanted stem cells of oligodendrocytes (cells from the central nervous system) into dogs with a demyelizing disease similar to human multiple sclerosis. They observed large areas of repair of the demyelinated area after transplant. Much work remains to be done before these results can be translated into human therapies, but the potential for successful treatment for multiple sclerosis is exciting.

Scientists initially believed that only early embryonic stem cells could become any type of cell in the body and that once stem cells had become localized to a specific organ, they could only change into cells specific to that organ. Recent research has shown that this view of the potentiality of stem cells may be too limited. Newer work is showing that adult stem cells that would once have been assumed to be committed to becoming specific mature cells can be reprogrammed to mature into an entirely different cell line. Scientists are also learning that they can program embryonic stem cells, those that possess all the potential to become any adult cell, into becoming exactly the type of cell needed.

Reporting in September 2001 in the *Proceedings of the National Academy of Science,* a group of researchers from the University of Wisconsin described their work with embryonic stem cells. They treated the embryonic stem cells with serum, the liquid component of blood, and found that they possess the expected characteristics of blood precursor cells. The researchers noted that this method could lead to both a better understanding of human blood formulation and perhaps, in the future, a novel way of obtaining blood cells for transfusion and transplantation.

Researchers at Yale University School of Medicine reported interesting findings on adult blood stem cells with seemingly unlimited potential in the May 2001 issue of *Cell.* To demonstrate the potential of these stem cells, the researchers irradiated female mice, killing off their own blood stem cells, and then transplanted male blood cells into them. By tracking the presence of the male Y chromosome, the scientists were able to demonstrate that the transplanted male stem cells had become not only bone marrow and blood cells, as expected, but also became lung, digestive system, liver, and skin cells.

The American Society for Cell Biology met in Washington DC in December 2001, and researchers presented data that showed that muscle stem cells had far greater potential than had previously been believed. The scientists reported that mouse muscle stem cells could migrate into the bone marrow of irradiated mice and repopulate the bone marrow with blood-making cells. Of great interest was the fact that when the new blood cells that had arisen from the muscle stem cells were harvested, they could revert back to producing muscle cells. Thus, stem cells are not as committed to producing one cell line as previously thought.

At the November 2001 meeting of the Society for Neuroscience, Dr. Lorraine Iacovetti presented findings from her laboratory at the Thomas Jefferson University Hospital. She and her associates focused on how stem cells could be made to switch from one type to another. They used a combination of different growth factors and nutrients and managed to convert a population of human blood stem cells into human neuron (nerve) stem cells. This work all had to be done in the laboratory, and the new neurons did revert back to their original form in just two to three days, but the groundwork is being laid for new research into the programming of stem cells into whatever cell lines are desired.

Although these results appear promising, scientists have not reached a consensus as to how adult stem cell reprogramming works. Uncovering the mechanisms of this new type of stem cell differentiation may provide additional insights into the use of stem cells for tissue repair.

Stem Cells and Aging

Most of the cells within the human body have defined life spans. Most cells can only divide and replace themselves between forty and fifty times. After a cell has undergone its fifty divisions, it can divide no further and enters into the cellular aging process. Thus, our organs have a finite life span, given that they are composed of cells whose function eventually declines. Stem cells, however, have unlimited potential to divide and virtually unlimited potential uses within the body. Scientists speculate that stem cells thus constitute a largely untapped "fountain of youth," which, if harnessed, could conceivably allow us to rejuvenate our tissue as needed, thus prolonging our lives.

A major consequence of aging is the aging of our tissues and the inability of our bodies to replace those aging tissues. Some of the diseases of aging to which researchers hope to apply stem cell technology include neurological disorders, including Parkinson's and Alzheimer's diseases, spinal cord damage, strokes, heart disease, diabetes, burns, and arthritis, both rheumatoid and osteoarthritis. A culture of a sufficient number of multipotent (able to change into many cell types) stem cells could conceivably be used in place of whole organs in transplantation.

Scientists have long used animal models to study human disease. Animal models also prove useful in the study of aging. Many of the cellular aspects of aging have been examined in animal models, and parallels have been found in human aging. One of the most studied animal models of aging is the roundworm, Caenorhabditis elegans.

Researchers at the University of California in San Francisco discovered that stem cells that are involved in the reproductive function of the roundworm also influence the aging process. The stem cells in question are not those that develop into eggs or sperm, but those that develop into cells that support the production of eggs or sperm. Removal of those cells, called proliferating germ-line stem cells, from the roundworm results in a dramatic increase in life span. The scientists suspect the germ-line stem cells interfere with a reproductive hormone that might otherwise promote longevity by either decreasing production of such a hormone or by inhibiting its action.

Huntington's Disease

Stem cell transplants may provide benefits in treating Huntington's disease according to a new study. Researchers found the benefits of experimental stem cell therapy in reducing symptoms, such as muscle stiffness and memory loss, peaked after two years and then faded four to six years after the procedure in people with Huntington's disease.

While not a cure for the rare and difficult-to-treat condition, the study suggests that transplanting healthy stem cells to replace those brain cells damaged by Huntington's disease may offer patients long-term improvements and stability. The progressive neurological disorder causes rapid, jerky movements, loss of memory, and behavioral problems.

In the study published in *The Lancet Neurology*, researchers followed five patients with Huntington's disease for up to six years following experimental treatment with stem cell transplants. Previously, a small pilot study showed the treatment led to improvements in movement and brain function for two years after surgery, but the long-term benefits of stem cell therapy in treating Huntington's disease were unclear. The results showed that the improvements reached a plateau two years after surgery and then faded away at a variable rate for four to six years after surgery. Two patients who did not benefit from stem cell transplantation two years after the procedure continued to decline in the same way as untreated patients.

Researchers say that further refinements in the stem cell transplantation technique as well as selecting patients appropriate for the procedure may improve the potential benefits of stem cell therapy in treating Huntington's disease.

Brain

Stunning new laboratory research suggests that some special cells in bone marrow could be used to repair and replenish the brain and central nervous system, offering potential therapies for stroke, brain injury, and Alzheimer's disease. In an overview of the emerging science of brain regeneration, presented at the recent meeting of the American Psychiatric Association, Dr. Darwin J. Prockop said the new revelations promise to catapult stem cell research into areas considered unimaginable just five or ten years ago.

Ultimately, stem cells from bone marrow with regenerative capabilities may be used as a veritable "fountain of youth" for the brain, says Dr. Prockop, who is director of the Center for Gene

Therapy at Tulane University School of Medicine in New Orleans. The wellspring of that fountain are special cells found in bone marrow known as marrow stromal cells, or MSCs, which appear to have the capacity to rejuvenate tissue in many parts of the body. Exactly how the cells work remains to be clarified. Some scientists believe they provide growth factors that encourage pre-existing stem cells in various parts of the body, including the brain, to grow anew and form other cells. Others, like Dr. Prockop, believe they act like stem cells, taking on the function of cells in whatever environment, including the brain or nervous system, to which they are introduced.

In any case, animal studies show that when MSCs are extracted from rats or mice, multiplied in the laboratory, then injected directly back into the animals' brains, remarkable things happen. "Some of the cells injected into the central nervous system become new nerve cells," Dr. Prockop says. "The cells really take off and migrate throughout the brain. Amazingly, we get more cells back than we put in."

"Some day," Dr. Prockop says, "the strategy might be extended to humans."

"Potentially, we can isolate MSCs from a patient, put the cells in a freezer, then infuse them into the brain of the same patient," Prockop says. "We could do this every two to five years as a renewable source of cells for the brain..."

MSCs can be easily extracted from bone marrow using a needle and syringe, a feature that distinguishes the strategy from the more typical therapeutic use of stem cells, which are commonly derived from animal or human fetal tissue. Still, the development of real treatments of MSCs is still many years away and much needs to be learned, Prockop says. Interestingly, MSCs work to repair the brain naturally, but they are too few in number and do it too slowly to keep up with the amount of damage that occurs in severe disorders or even in the amount of brain cells that are lost naturally.

But when extracted from bone marrow and grown in a laboratory, MSCs can be produced in far higher numbers than exist naturally in the body. "In the body they are growing very slowly, but when we remove them from their natural environment and grow them in the laboratory, they explode. We can produce 10 trillion cells in six to eight weeks," Dr. Prockop says.

Regardless of whether MSCs activate existing stem cells in the brain or act like stem cells themselves, Prockop says the discovery that stem cells exist in the brain, serving as a natural reservoir of repair, is itself a revelation. It overturns the long-standing accepted wisdom that people are born with a fixed number of brain cells that are irrevocably lost when the cells die.

"We used to believe that the adult brain did not have stem cells," Dr. Prockop says. "As of about five or ten years ago, that has been thoroughly disproved. There are stem cells that are self-renewing and they are capable of repairing the brain. In effect, they are cells that are slowly trying to replace tissue and cells. But with MSCs we can speed up the process." Experts in the field agree that real treatments are years away, but the use of marrow stromal cells to repair and replenish the nervous system is not science fiction.

"These cells are very intelligent," says Michael Chopp, PhD. They target specific areas of the brain and behave like little factories producing an array of factors, which cause remodeling of the brain at very early stages after being introduced. Dr. Chopp is vice chair of neurology at Henry Ford Hospital, in Detroit, and professor of physics at Oakland University in Rochester, Michigan.

Dr. Chopp believes MSCs work primarily by activating pre-existing stem cells in the brain to provide an environment in which damaged cells can rejuvenate. "They are providing all the factors to promote self-repair," he says. "They jump-start the brain." In a recent report in the medical journal *Stroke,* Dr. Chopp and colleagues described how they induced stroke in rats to test the hypothesis that injection of MSCs could enter the brain and reduce neurological damage.

What they found was striking. The rats injected with MSCs showed significant recovery of function and improvement one week after injection, compared to the rats who did not receive the MSCs. Moreover, MSCs survive and take up residence in critical areas of the brain, with a few of them actually behaving like new nerve cells. Within days or weeks after treatment, the animals behave much better. So far-reaching and radical is the potential promise of MSCs that even scientists involved in exploring the possibilities are awestruck.

"I would have been embarrassed to speak about this two or three years ago," Dr. Prockop says. "I didn't believe it until very recently. I cannot prove that we can use the marrow as the equivalent of a fountain of youth, but we are moving close to that possibility. Now is the time for us to seriously debate it and discuss the consequences of where the science is moving. We need to discuss this not only among scientists but also among physicians and the lay public."

Shape-Shifting Cells Can Become Bone, Cartilage, Muscle, or Marrow

One day, doctors may treat bone loss with cells harvested from your arteries. It could happen now that researchers have discovered a new population of stem cells. The finding explains why for centuries butchers have sometimes found bone growing inside blood vessels. "It also explains hardening of the arteries," says Linda L. Demer, MD, PhD. Demer led the UCLA team that discovered what the cells do.

"The reason that arteries harden is that bone is growing in them," says Demer. Demer's team previously showed that this bone doesn't come from somewhere else in the body. They traced the bone growth to calcifying vascular cells, or CVCs, found in artery walls and heart valves. Unlike normal cells, CVCs can turn into other types of cells.

"We have now shown clearly that CVCs not only become bone, but cartilage, marrow, and smooth muscle as well," Dr. Demer says. The new findings appear in the October 2006 early access issue of *Circulation: Journal of the American Heart Association.*

And they have an extra trick: they can self-renew, thus representing a lifelong source of new bone and cartilage cells.

On the other hand, CVCs can't turn into fat cells. But that may be a big advantage if the cells can be harvested, grown to larger numbers, and used to treat disease. "Older women who are losing bone in their skeletons are still forming new bone in their arteries," Dr. Demer says. That suggests that they still have the capacity to create bone. It may be possible to harness that process to create a therapy for bone loss.

Knee Repair

Osiris Therapeutics, Inc. has completed enrollment in the first clinical trials of a stem cell therapy to repair tissue in the knee. A total of fifty-five patients were treated in the Phase I/II double-blinded, placebo-controlled trial to evaluate the safety and effectiveness of Chondrogen, a preparation of adult stem cells formulated for direct injection into the knee.

"Over the past 15 years there has been much laboratory research demonstrating the ability of these stem cells to regenerate orthopedic tissues. It is highly gratifying to now be able to move the science from the laboratory into the clinic, where the cells can be evaluated in humans," said lead investigator C. Thomas Vangsness, MD, professor of orthopedic surgery at the University of California's Keck School of Medicine and chief of sports medicine at University Hospital.

In the U.S. alone, approximately 800,000 people each year have surgery to remove a damaged or torn meniscus, a cartilage-like tissue in the knee that acts as a shock absorber. While the surgery often relieves pain associated with a torn meniscus, patients who have this surgery are at a much greater risk of developing arthritis. Chondrogen is being studied to regenerate the meniscus in these patients.

"This study is the third clinical trial for which we have completed enrollment this year," announced C. Randall Mills, PhD, president and CEO of Osiris, "...and I would especially like to thank the patients participating in this trial, as well as other clinical trials. It is their willingness to participate that contributes to advancements in medicine."

"There is currently no treatment available to regenerate a damaged meniscus. The only option for these patients is to remove the damaged tissue," said Joel Boyd, MD, an orthopedic surgeon at TRIA Research Institute in Minneapolis, Minnesota, and a U.S. Olympic Team physician. "The ability to give patients a simple injection into the knee that could restore the meniscus and prevent the inevitable progression to osteoarthritis would be a significant advancement in the treatment of knee pain."

Breast Tissue

In Japan, Cytori Therapeutics, Inc. has begun its first clinical trial to test whether stem cells derived from fat can be used to regenerate breast tissue in women. A twenty-patient study, which includes women who have undergone partial mastectomies due to breast cancer, was approved by Japanese regulators. Cytori, based in San Diego, said it has developed a system for isolating and concentrating a patient's own stem cells from tissue removed from one of the body's normal fatty deposits.

Tendons and Ligaments

A new approach for tendon regeneration was reported by the *Journal of Clinical Investigation.*

Israeli researchers are using adult stem cells to regenerate torn tendon and ligament tissue at Hebrew University in Jerusalem. Adult stem cells are being used by an Israeli research team to create a new orthopedic solution to the difficult and common problem of how to heal torn ligaments and tendons.

The research team, led by Professor Dan Gazit, is working to change this by using stem cells taken from bone marrow and genetically engineering them to become different cells altogether. "With this in mind, we can genetically engineer new skeletal tissue such as ligaments, cartilage and tendons," explained Dr. Gadi Peled, a senior scientist at the lab.

The stem cells were injected with two proteins, Smad8 and BMP2, and were then injected into the torn Achilles tendons of rats at the Skeletal Biotechnology Laboratory at Hebrew University. The cells were drawn to the site of the injury and were able to repair the tendon. There was complete healing in seven weeks. A special type of imaging test was used in the study to identify the tendon tissue repairs. The test, known as proton DGF MRI, differentiates between tendons, bones, skin, and muscle, all of which contain different amounts of collagen.

Dr. Peled said that in time, the technology might be used to help people who suffer with lower back pain as well because invertebrate discs consist largely of tendon tissue which deteriorates over time. The team's next step will be to conduct the tests on large animals. Goats are the best subjects. Pigs will be used for invertebrate disc research.

Clinical trials on bone regeneration are currently being conducted on humans by Hadassah Medical Center together with Teca Pharmaceuticals, using the same adult stem cells.

Chapter Eleven

Take Charge of Your Own Case

Our doctors do everything within their power to ease our suffering, but they are limited to therapies that the Food and Drug Administration (FDA) has approved for general use. Surprisingly, physicians are not likely to know about all of the advanced treatments that are in the clinical trial phase in hospitals and laboratories around the world. Although they keep up with approved new therapies by reading medical journals and participating in required continuing education courses, their busy practices keep them from spending the large amount of time necessary to be aware of all advanced treatments not yet approved for general use. If an illness occurs that requires therapy beyond the scope of what your doctor or a local specialist can provide, several situations can arise between you and your doctor.

I am fortunate to have a cardiologist who did everything possible to refer me to other doctors and medical facilities. He wrote referral letters and sent medical records to every possible source of help that either he or I could identify. I hope you have a doctor with the same concern for your well-being. However, this may not always be the case.

The following is an example of a situation that could arise between you and your family doctor. Your doctor has treated you and your family for many years. You are very pleased with his care and completely trust in his judgment. He has handled your family's illnesses very well, and you consider him a friend as well as a doctor. However, a new condition has arisen. Your husband is very sick with an illness that is nearly always fatal, and he seems to be getting progressively worse. Your doctor referred him to a specialist, but your husband did not respond to any conventional therapies. Nothing helped, and you will never forget these dreadful words uttered by your doctor: "I'm sorry, but that's all we can do for him." You and your husband try to reconcile yourselves to the terrifying news that there is no hope and that you must plan for the inevitable.

A few days later, you speak with your best friend about your husband's illness, and she tells you of an acquaintance with the same condition. The acquaintance found out about a clinical trial of a new stem cell therapy that a highly respected medical institution was conducting. She said that the Food and Drug Administration (FDA) had approved the therapy for the clinical trial, but more testing was needed before it could be approved for use by doctors in general practice.

Your daughter, who is very familiar with computers and the Internet, conducted a search and found that, sure enough, a clinical trial using stem cell therapy to treat your husband's illness is underway at a nearby major medical institution. She also found that, in many cases, stem cells extracted from the blood of the patient could repair damaged organs and improve function. Your daughter immediately called the trial coordinator who confirmed the Internet information and asked some questions about your husband's illness. She said that the trial was still seeking participants and that your husband could possibly be a candidate for the new therapy. The trial coordinator requested that your husband's doctor send your husband's complete medical records for evaluation by the doctors on the clinical trial team to determine if he meets the requirements for inclusion in the trial.

You are ecstatic. Here is hope for your husband's recovery! You understand that there are risks associated with any therapy, but nothing else is available to your husband. This could save his life. You and your husband immediately make an appointment with your doctor to tell him the great news. After he listens to your story, your doctor completely shocks you when he says that he has been practicing medicine for thirty years, has heard many wild stories about stem cell therapy, and feels that it could not possibly be helpful. He tells you that scientists have done some testing in laboratories, but that such therapy is still years away from human use. "No, I'm afraid I can't go along with this" are the next words you hear from the doctor.

Your husband responds with, "Well, maybe we should get another opinion about this. Who would you recommend?"

Then the next shock comes. Getting red in the face, your doctor says, "I've treated both of you for many years and given you the best that medicine can provide. If my work hasn't been good enough for you, perhaps you should replace me with someone else. Let me know who you find, and I'll send your records to his office."

What do you do now? Most likely, it would be appropriate to talk further with your doctor to discover what is behind his strong feelings. There are several possible reasons he feels the way he does. In this example, it may have to do with the source of the stem cells themselves. Because he thinks that stem cells could possibly have come from a hepatitis or HIV victim, he thinks your husband could somehow become infected. He does not have sufficient information to know that use of your husband's own stem cells makes this impossible. On the other hand, your doctor may have solid information from authoritative sources that the proposed stem cell clinical trial offers no benefits and very high risk.

Chances are your own doctor will not feel insulted because you suggested a second opinion. All doctors know they could be wrong, and they sometimes suggest that you get another opinion. This has benefits for him as well as for you. For one thing, it gives doctors the opportunity of getting their work checked by another qualified doctor. The second doctor could completely agree with your doctor and may suggest alternative treatments, or the second doctor may disagree with your doctor entirely. In which case, the doctors can decide together on a treatment plan in your best interest.

Another possible cause for the doctor's reaction, which could especially be true of older doctors, is that the nature of patients themselves has changed with the advent of the Internet and with the huge bombardment of pharmaceutical advertisements on television.

Not too many years ago, the average patient regarded physicians as supreme authorities. He or she believed everything doctors had to say. When given treatment plans, they followed "doctor's orders" because they had a fear of terrible consequences if they failed to comply. Then, the Internet came along. Soon, many patients were jumping onto the World Wide Web and learning about their diseases even before going to their doctor's office. They found the names of diseases, causes, and possible courses of treatment. When visiting the doctor, instead of asking "What do I have? What can you do?" many patients began asking, "Do you think I should do this? Do you think I should do that?" Pharmaceutical companies, using a barrage of commercials on television, suggest that patients ask their doctors if new drug "X" is right for them. The doctor is put into the position of responding to the patients' preconceived ideas and in many cases, informing them that the information they got from the Internet is just plain wrong or that the drug advertised on television is not right for them because of their personal medical conditions. The older doctor may resent being put into this position. He has spent many years in medical school and many years in practice, and suddenly, patients have access to the most current medical information. This is especially true with regard to gene and stem cell therapies that are arising at an explosive rate. Your doctor stays informed of therapies approved by the FDA by reading medical journals and through continuing education courses, but it is almost impossible for him to maintain an active practice and to be aware of all of the stem cell clinical trials around the world.

Getting back to the dilemma with your doctor, there is no correct answer for everyone. However, one thing is certain: you must take charge of your own case. It is up to you or your loved ones to become informed and to make decisions that can often be critical. Even if you do nothing at all, you are taking charge of your own case because it is your decision to do nothing.

In many cases, you are fighting for your life. Do the best you can, even if you are tired and worn down at this critical time. Don't let your precious life slip away without the best fight you can muster. I was told, in confidence by a clinical trial coordinator, about an unnamed patient who was selected by a clinical trial team for a therapy that could save his life. He was very enthusiastic at first, but he checked with his local doctor, who said it was not a good idea. The patient turned down the opportunity, telling the trial coordinator that he planned to move to Florida and wait for his death. In your present condition, you may feel too tired of suffering to want your life to go on. However, please think ahead. If your suffering is relieved, you probably will think differently and be very happy that you made the decision to take advantage of every life-saving possibility.

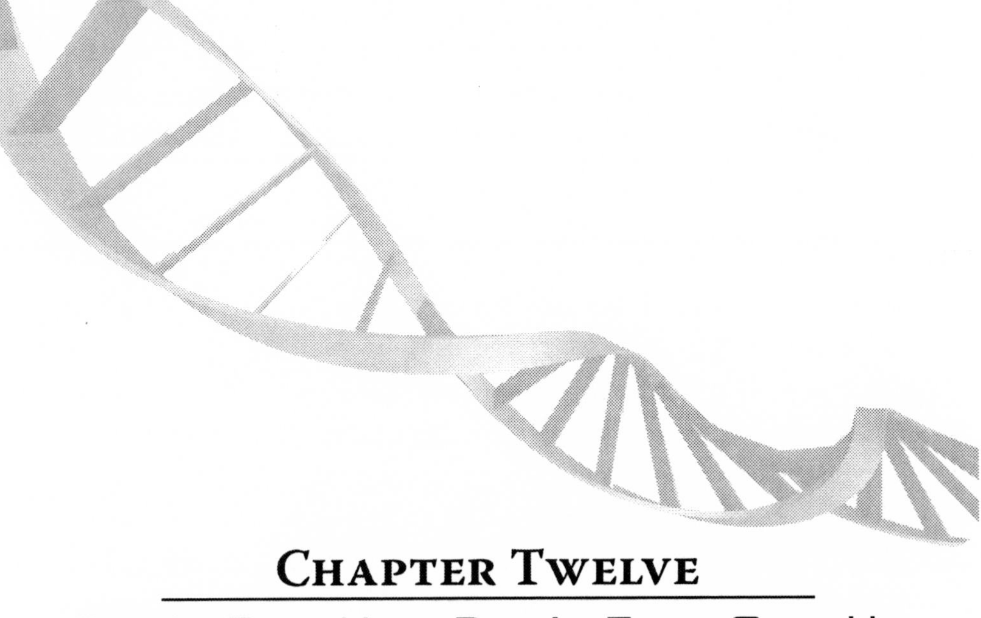

Chapter Twelve

Don't Give Up—Don't Ever Give Up

These stirring words are attributed to Jim Valvano, a famous basketball broadcaster who successfully fought the ravages of cancer. Similarly, during the darkest days of World War II, Winston Churchill said it another way, "Never give up; never give in."

These encouragements could not be more important to those who are fighting serious disease.

One of America's preeminent heart specialists, Dr. Michael E. DeBakey, says in his book *The New Living Heart* that after the heart is damaged, the aftermath can be difficult if, like most, you do not recover completely. Anxiety after a heart attack is replaced, in many cases, by depression. Confidence in knowing recovery is possible is most important to the recovery progress itself.

It is important to do everything possible to maintain or improve your condition by taking care of yourself as best you can. Strictly follow your doctor's advice on joining a cardiac recovery program or other follow-up regimen specific to your disease. Stick to the diet suggested by your doctor, exercise to the extent permitted by your doctor. Take your medications faithfully. Joining a support group is important no matter the nature of your illness. For heart patients, Mended Heart, Inc. offers encouragement and hope for heart patients, their families, and caregivers. There are support groups for almost every major illness. You can locate them on the Internet.

Books, such as *The Power of Positive Thinking* by Norman Vincent Peale, provide tools for forming a positive outlook. There are many other books available in the library, online, and in retail book stores.

It may seem that your doctor should know of all of the medical centers where advanced treatments are being studied and clinical trials may be available. However, this is probably not the case, because your doctor is very busy seeing many patients like yourself. Most likely, he or she reads journals concerning new therapies that have been FDA approved and are available to your doctor for his practice. It is unlikely, however, that your doctor has time to do an exhaustive literature search to find information for you about newer treatments in the laboratory and in clinical trials, so you must take charge of your own case and search out the necessary information.

Hold off on doing your literature study on advanced treatments and clinical trials until you have a positive outlook and are fully able to concentrate on what you are doing. If you are not well enough to pursue this task on your own, ask a friend, relative, or caregiver to do the study for you. Because the immeasurable amount of information about clinical trial availability can be confusing, a service is available on the Internet at clinicaltriallocator.org that can do an exhaustive search for you to find appropriate clinical trials for your disease.

When you perform the Internet study, follow every reference or link to additional information. What you are seeking may require following several links leading from one source to another.

The search may take a great deal of time and effort to find a helpful advanced treatment or clinical trial. Don't give up—don't ever give up.

Appendices

Appendix A

Sample Informed Consent Form

Caritas St. Elizabeth's Medical Center

Informed Consent For Experimental Procedure
(REVISED 3/28/03) (Retyped from Original)

Subject's Name: ______________________________

Date: ______________________________

Home Address: ______________________________

Home Telephone: ______________ Date of Birth: ______________

Research Study Title: Injection of Autologous CD34-Positive Cells for Neovascularization and Symptom Relief in Patients with Myocardial Ischemia

Principal Investigator: Douglas W. Losordo M.D.

Study Sponsor(s) Douglas W. Losordo, M.D., Caritas St. Elizabeth's Medicals Center, Boston, MA

The purpose of this consent form is to inform you about the nature of the Research Study so that you may make an informed decision as to whether you would like to participate. You are free to decline participation and, should you choose to participate, you are free to withdraw from the Study at any time without penalty or loss of benefits that you otherwise enjoy outside of the Research Study.

1. Invitation: You are being asked to participate in a research study. Your participation is voluntary.

You are being invited to participate in a research study to look at the safety and possible activity of an experimental cell therapy for patients with poor blood flow to areas of the heart. This new procedure is an investigational therapy that is highly experimental, not demonstrated to have a proven benefit and, not approved for use in the general public.

Your doctor has determined that you have coronary heart disease with a decreased blood supply to areas of your heart. Therefore, you may be eligible for participation in this research study.

2. Purpose: What is the purpose of this research study?

The goal of this study is to determine the safety of various doses of *autologous* (one's own) stem cells, delivered with a catheter into the regions of the heart with poor blood flow. Stem cells are primitive cells produced by bone marrow that can develop into blood cells or other types of cells. In addition to determining whether this approach is safe, the diagnostic tests may offer preliminary insights into the usefulness of this approach for treating *myocardial ischemia* (the condition where areas of the heart are lacking enough oxygen and blood flow to keep the heart muscle working well).

This is a blinded, randomized study to compare a certain type of stem cell called a CD34-positive versus a placebo agent (normal saline). *Randomization* is a way of randomly assigning treatment, similar to picking numbers out of a hat. You will have a 3:1 chance of receiving your CD34-positive stem cells versus the placebo agent (normal saline). *Blinded* means that neither you nor the Caritas St. Elizabeth's Medical Center research team will know whether you received your CD34-positive stem cells or the placebo agent. If you are randomized to receive placebo (normal saline), you will undergo all of the pretreatment phases of this study (including the stem cell mobilization phase and the apheresis procedure), but rather than receiving injections of CD34-positive cells, you will receive injections of the placebo agent (normal saline).

Any additional cells that are collected from you will be stored (frozen). After you have completed at least 6 months of follow up, and everyone in the dose group you are assigned have completed at least 3 months of follow up, if you have received the placebo and are still meet [sic] the study entry criteria, you will be given the option of receiving your cells. If there are not enough frozen cells, or for any other reason they do not pass the required testing, you will be given the option to go through the process of having the cells mobilized and collected again prior to having them injected into your heart. There is some research evidence that CD34-positive cells may develop or improve blood flow when injected directly into the heart muscle.

3. Duration: How long will you be participating in this research study?

The approximate duration of your participation in this study is 12-13 months. If you initially received placebo, and after at least 6 months you still meet the study criteria, and choose to be treated with your own cells at that time, this would extend the duration of your participation in this study to 18-20 months (6-12 months in the placebo group and 12 months in the treatment group). It is extremely important that you maintain your follow up schedule as part of your participation in this experimental protocol. It is also necessary that you complete all follow-up testing at Caritas St. Elizabeth's Medical Center.

4. Procedures: What will the research study involve?

You understand that you will visit the clinic, before receiving cell therapy to determine if you are eligible to participate in the study. These visits will be scheduled up to 8 weeks before treatment.

SCREENING VISIT

Your doctor will ask you detailed questions about your current health, medications and medical history. Your doctor will perform a physical examination and measure your vital signs (blood pressure, heart rate, pulse and temperature). You will be given an electrocardiogram or EKG (measures heart beats), an echocardiogram (an ultrasound picture that shows how your heart muscle is functioning) and a chest x-ray. You will have approximately 5 tablespoons of blood draw (hematology and chemistry) and a urine sample will be given. Your blood will be tested for infectious diseases including Hepatitis B, Hepatitis C and HIV. Woman of childbearing potential will have a serum pregnancy test performed. You will fill out a questionnaire about how you are feeling and what activities you are able to do. The veins in your arms will be checked, similar to before you donate blood; to make sure you can have blood collected easily and returned to you during the special procedure to collect your blood cells. If your veins won't easily allow a needle and tube to be inserted, you may have to undergo a procedure to have a catheter (or tube) inserted elsewhere (such as your chest or neck) for blood collection process. This will only be done if needed, if you qualify for the study.

To determine if you are eligible to participate in this study you will have an exercise test on a treadmill and a nuclear stress test. The nuclear stress test will give your doctor information about which parts of your heart are healthy or not. During this test you will be injected with a very small amount of a radioactive substance. Once in the body, this substance emits rays that can be picked up by a special (gamma) camera. The rays allow the camera to produce clear pictures on a monitor of your heart tissue. These pictures show contrasts between light and dark spots, which can indicate areas of damage or reduced blood flow that are present before, during and after stress. For this test, you will be given a medication called adenosine in a vein in your arm (intravenous or i.v.), which will have a similar effect to when you exercise on the treadmill – your heart rate will rise and your heart muscles will work harder. At the end of the stress part of the test, you will receive another small amount of a radioactive tracer and then another set of pictures will be taken of your heart with the special camera after your hart rate has returned to normal (rest).

In addition, your measurements will be taken for a special wearable defibrillator vest, called the LifeVest, which you will wear to monitor your heart rhythms after you have received the study treatment, if you meet the study requirements. The LifeVest will signal you by vibrating and by sound before delivering the shock, so, you are feeling ok, you can stop the shock from coming. The information about your heart rhythms will be sent to a computer once each week so your doctor can review this information.

Stem Cell Mobilization Phase

After you complete the screening and you are identified as a candidate, you will begin a process to help move some of your stem cells from your bone marrow into your blood circulation. You will be given a drug called G-CSF or Neuopogen® (a drug used to move cells from the bone marrow into the bloodstream). **If you have an allergy to E coli-derived proteins you may not take this drug or participate in this study**. You will receive a daily injection of G-CSF for up to five days. Blood samples will be drawn every day to monitor the effect of G-CSF on moving cells into the bloodstream. You will be scheduled to donate your own blood samples for treatment in the study on the day you receive your last dose of G-CSF.

On the day of your first injection of G-CSF, you will be shown how to use the LifeVest wearable defibrillator, which you will wear day and night, except when bathing, for the next five weeks.

Apheresis

After receiving your last dose of G-CSF, stem cells will then be collected by a procedure called apheresis. Apheresis is a method of collecting large numbers of certain blood cells, such as white cells, stem cells, or platelets. In this procedure, whole blood is collected through a needle in an arm vein, similar to donating blood. You will lie in a recliner and a needle is placed in a vein in each arm. Blood is collected from one arm and passes through a special machine called a blood cell separator. The machine collects the stem cells in a bag and directs the rest of the blood back through the needle in the other arm. If your veins won't easily allow a needle and tube to be inserted, you will need a catheter (or tube) inserted elsewhere (such as in your chest or neck) for the blood collection process. In this case, the blood is collected and returned through the same catheter. One-half tablespoon of blood is drawn before and after each donation to measure blood cell counts. You will be monitored closely during the procedure. Any additional cells that are collected from you will be stored (frozen).

Treatment

You will be given 10 tiny injections (4/100ths of a teaspoon for each injection) of your own cells directly into areas of the heart that are not getting enough blood flow. You will either receive specific type of cell collected from within your blood called CD34-positive cells, or, you will receive placebo of salt water (normal saline). Even though the injections are very small, each injection contains between one hundred and several thousand stem cells. This study is blinded – you will not know whether you received CD34-positive cells or the placebo agent (normal saline). There will be three groups of patients in the study. Each group will have eight patients, 6 of whom will receive one dose, or amount, of cells, and two of whom will placebo agent (normal saline); the first group of patients to assigned to receive cells will receive 5×10^4 cells/kg body weight, the subsequent group 1×10^5 cells/kg body weight, and the last group 5×10^5 cells/kg. The total amount of cells in each group is divided up into 10 injections. The patients assigned to receive the placebo agent (normal saline), will also receive 10 injections of the placebo agent (normal saline).

The injections are delivered with a special catheter system that is inserted with a small incision into a blood vessel in your groin. This catheter system is experimental and only available for use in this study; it is not available for use by the general public. You will have local anesthesia at the site where the catheter is inserted but you will not typically have general anesthesia. You will undergo a cardiac catheterization with a special catheter inserted into the left ventricle of your heart. This special catheter is moved over the inside surface of your heart and provides information on a monitor showing electrical activity and motion of your heart. This information will be used to identify areas that are scarred from a previous heart attack as well as other areas that have limited movement due to reduced blood flow. This will help your doctor determine where to inject your stem cells, or the placebo agent (normal saline) into the heart muscle. The needle injection catheter that is part of this system will then be used to deliver your cells or the placebo agent (normal saline) into these areas of the heart. After the injections are performed, the catheter is removed and the incision in your groin is closed. You will have to remain flat until the next day. Your heart and vital signs will be monitored for the next 24 hours or longer until your doctor decides you are well enough to leave the hospital.

Post-treatment monitoring

After the procedure, you will be monitored closely for the next 24 hours. Your vital signs will be checked every hour and you will have your heart monitored. When your doctor is ready to discharge you from the hospital, you will have a physical exam, measure your vital signs and review your medications. An echocardiogram and an EKG will be performed to look at how your heart is functioning. You will also have approximately 5 tablespoons of blood drawn (hematology and chemistry) and a urine sample will be given. Before leaving the hospital, you will again be instructed on how to use the LifeVest wearable defibrillator, which you will continue to wear day and night, except when bathing, for the next four weeks.

Follow-up Visits

After you are discharged from the hospital following your procedure, you will return to the clinic in 1 week and again at 1,2,3,6,9 and 12 months to see how you are doing. At every visit, your doctor will ask you detailed questions about your current health and medications. Your doctor will perform a compete physical examination and measure your vital signs (blood pressure, heart rate, pulse and temperature) You will be given an EKG at each visit and an echocardiogram at 1 week and 1,3 and 6 months. At each visit, (with the exception of 2 months) you will have approximately 5 tablespoons of blood drawn (hematology and chemistry) and a urine sample will be given. If you are not wearing the LifeVest wearable defibrillator, you will have a 24 hour holter monitor test (wearing EKG leads and a monitor) at 1 week done to check how your heart is beating over the course of a day and night. After 4 weeks, you may stop wearing the LifeVest. You will have additional holter monitor tests at 3 and 6 months. You will also fill out a questionnaire about how you are feeling and what activities you are able to do at 3, 6 and 12 month visits. When you return for the 3 and 12 month visits, you will have another exercise treadmill test. You will also have follow-up nuclear stress scan done at 3 and 6 months.

If you are randomized to receive the placebo agent, you will have the opportunity to cross-over into the treatment group. At your six-month visit you will be evaluated for the cross-over if you received the placebo agent. You will need to continue to meet the study criteria you did at the baseline for re-entry into the treatment group. If you meet the study criteria you may receive your cells that were collected by apheresis during the baseline study phase. If your CD34-positive cells, previously collected and frozen, are unable to be used, you will have to undergo the stem cell mobilization phase and apheresis again. If you choose to cross-over into the treatment group, this will extend your study time up to 24 months (6-12 months in the placebo group, and an additional 12 month in the treatment group). If you qualify for the cross-over, and your request for cross-over is granted, you will be dosed with the current active dose regardless of your original dosing group.

5. Risks, Discomforts, Side-effects and Inconveniences: **What are the risks involved with being enrolled in this study?** During each stage of the study protocol, there are some possible risks and side effects you might experience.

Pre-treatment and study testing: During the nuclear stress test and during the exercise testing, you may experience some feelings of chest pain and/or shortness of breath you experience at other times when you do certain activities. Your heart rate and blood pressure will be monitored throughout these tests and the exercise or drug caused stress to your heart will be stopped if you experience chest pain or have any other problems. When you have nuclear scans, you may experience slight discomfort from the needle piercing the skin, when the radioactive substance is injected twice. You could also have an allergic reaction, including but not limited to; a minor reaction of rash, hives or a more severe reaction of trouble breathing and shock, due to the drug given for stress.

Potential risks of stem cell mobilization The substance that you will be taking to move your stem cells from your bone marrow into your circulation may cause side effects. G-CSF or Neupogen® is contraindicated in patients with known hypersensitivity to *E-Coli*-derived proteins, Filgrastim, or any component of the product. Allergic-type reactions occurring in initial or subsequent treatment have been reported in <1 in 4000 patients treated with Neupogen®. Allergic reactions include rash, swelling of your face or body, difficulty breathing, low blood pressure and irregular fast heart rhythm. Some reactions occurred on initial exposure. Other potential side effects include increased frequency or intensity of angina, bone pain, headache, and changes in your blood counts or anemia.

Potential risks of apheresis During the stem cell collection process, you may feel tingling around your mouth, hands or feet. This is due to a blood thinner in the collection machine, which is added to the blood to prevent the blood from clogging in the apheresis machine. Your vital signs will be monitored carefully throughout the donation process and the procedure will be stopped at any time if you are not feeling well. Bleeding, infection, return of air or blood clots into the bloodstream, allergic reactions, and extremely rarely, cardiac arrest and death are potential risks of the collection process.

Potential risks of stem cell or placebo injection into the heart As with any new drug, an allergic reaction could occur. This could result in hives, rash, difficulty breathing or collapse of the lung and breathing systems. These reactions are usually reversible but could lead to permanent disability or death. Other rare but possible risks include infection, respiratory distress or failure, pulmonary edema, renal or hepatic failure. As with any investigational drug, administration of autologous CD34-positive cells could cause currently unknown side effects that might be permanently disabling or fatal.

Cell therapy will be delivered percutaneously with a retractable injection catheter. Possible risks associated with the delivery are minimized due to the nature of the study agent, namely autologous CD34-positive cells. It is possible that a small amount of the cells will enter the systemic circulation of the heart rather than the myocardium, although since these are autologous cells that are normally present in the circulation there should be no sequelae of such an event. If the delivery catheter were to penetrate the heart (from the inside to the outside) there is the possibility of adverse effect from the cells appearing in the pericardial space surrounding the heart, potentially causing an inflammatory response. Injection directly into the myocardium also may cause inflammation or halter recordings beginning immediately after cell therapy delivery. Patients randomized to placebo will have the placebo (normal saline) injected into the heart muscle instead of CD34-positive cells. These patients are subjected to the same risks of the injection procedure as patients receiving CD34-positive cells, and will be closely monitored in an identical fashion to those patients receiving the CD34-positive cells.

Potential risks associated with cardiac catheterization

This protocol requires insertion of a catheter to inject the autologous selected or non-selected cells into the myocardium. This method has been employed in previous studies to analyze treatment effect of other drugs designed to improve myocardial perfusion. The risk of significant adverse events from cardiac catheterization is small, but could include arterial perforation or dissection, hematoma or hemorrhage, coronary artery spasm or embolism, significant brady- or tachyarrhythmias, myocardial infarction, cerebrovascular event, and death. There is a small risk that the needle injection could go through the heart and cause the cells or placebo (normal saline) to get into the space around your heart (pericardium) or cause cardiac tampenade.

Potential risk associated with the LifeVest wearable defibrillator As a safety measure, you will be given a LifeVest to wear for the first month following study treatment. The LifeVest is an automatic wearable defibrillator which will charge and deliver a shock to your chest if your heart develops a very rapid heart rhythm. There is a small chance that the LifeVest could deliver a shock after your heart rhythm has returned to normal on its own, if you are not able to or do not cancel the shock. It is also possible, in rare situations for a shock to be delivered when you are not experiencing an abnormal heartbeat. The alarms on your LifeVest will alert you to the device charging and if this happens, you can also prevent the shock from being delivered. If a shock is delivered, you may experience redness or burn to your skin where the shocking electrodes are placed.

Potential risks of being crossed over from placebo to active treatment If you are randomized to placebo treatment, and choose to receive your own cells after you have completed at least 6 months of follow up and all subjects in your assigned dose cohort have completed at least 3 months of follow up, you will potentially experience the same risks as all of those stated above. There may also be currently unknown side effects that might be permanently disabling or fatal.

6. Benefits: **Are there any benefits to participating in this study?**

You understand that it is not possible to predict or guarantee positive response to this experimental treatment. It is currently unknown whether injected some of your own cells into the heart will cause new blood vessels to grow and it is not known what dose or amount of cells is needed to grow new blood vessels. It is hoped that stem cell transfer will improve blood flow in areas of your heart, although that benefit is by no means proven. If there is improvement in blood flow in your heart, this result may improve your ability to do more activities before experiencing chest pain. The information gained in this study may improve the understanding of the safety of using stem cells in the heart and aid in advancement of medical knowledge.

7. Alternatives: **Therapy is available to you without enrolling in this study. The appropriate alternative procedures or courses of treatment include the Following:**

You have coronary heart disease with reduced blood flow to areas within your heart. Alternate therapy includes the current treatment you receive from your doctor, including medicines, diet and nutrition. You may or may not be able to have other treatments at this time due to the seriousness of your heart disease. It may be possible for you to undergo additional revascularization procedures such as angioplasty and stenting or surgery, including coronary artery bypass grafting.

8. Confidentiality:

Confidential information contained in your medical record may not be given to anyone except to members of the research group and others who must be involved professionally to provide essential medical care. The study sponsor, the Research/Human Subjects Committee (IRB), and the federal agencies including but not limited to the Food and Drug Administration, protecting the welfare of study participants may view the records.

9. Compensation: **Will you be paid to participate in this research study?**
   - ( ) You will be compensated for participating in this research study. You will receive:
   - ( ) You will not receive any sort of compensation for participating in this research study

10. In Case of Injury:

If you become sick or injured in your direct participation in this research study, medical treatment will be provided to you including first aid, emergency treatment and follow up care as needed. Caritas St. Elizabeth's Medical Center will bill you health insurance for the cost of such care. If your insurance does not pay for your care, or pays only a portion of the cost of such care, Caritas St. Elizabeth's Medical Center may bill you for any unpaid amounts. No special arrangements will be made for the compensation or for the payment of treatment solely because of your participation in this research study. Caritas St. Elizabeth's Medical Center and persons conducting this research study are not admitting fault for you injury or illness by providing or making available medical treatment for your injuries or illness. This paragraph is a statement of the Caritas St. Elizabeth's Medical Center Policy and does not waive any of your legal rights.

11. Costs: **What charges will be paid by the study sponsor?**

You understand that you may have to pay for medicines, devices, and other medical supplies and services not related to your participation in the Research Study. Caritas St. Elizabeth's Medical Center will first bill your health insurer for such costs. If your insurance does not pay, or pays only a portion of the costs, Caritas St. Elizabeth's Medical Center may bill you for any unpaid amounts.

12. New Findings: **New Information**

Any new findings developed during the course of this Research Study, which may affect our willingness to continue participating will be explained to you and you can then decide if you want to continue in this research study. (and your consent to continued participation will be required)

13. Number of Subjects:

The number of subjects who will participate in the Research Study at Caritas St. Elizabeth's Medical Center is estimated to be 24 in this pilot study.

14. Termination without Consent:

You understand that Dr. Losordo may terminate your participation in this study at any time. Specific reasons for discontinuation include death, myocardial infarction (heart attack), recurrent angina, revascularization of a coronary vessel, or coronary artery bypass graft. You will be terminated from the study if your physicians deem the interventional procedure not successful and you require another type of treatment at that time. You will, however, continue to be monitored for safety but no further follow-up interventions will be performed.

15. Contacts:

If at any time during this research study, you feel that you have not been adequately informed as to the risks, benefits, alternative procedures, or your rights as a research subject, or feel under duress to participate against your wishes, you can contact a member of the Research/Human Subjects Committee, who will be available to speak with you during normal working hours (8:30 a.m. to 5:00 p.m.) at:

Institutional Review board (IRB) Research/Human subjects Committee
Telephone:
Address:

You may also contact the Principal Investigator, Dr. Douglas W. Losordo, at any time during this Research Study for questions and answers regarding the Research Study at (XXX) XXX-XXXX, Dr. Peter Soukas at (XXX) XXX-XXXX, Dr. Pinak Shah at (XXX) XXX-XXXX or the on-call cardiologist at Caritas St. Elizabeth's Medical Center (XXX) XXX-XXXX. If you feel you have been injured as a direct result of this study, please contact Dr. Losordo. In the event that additional experts, particularly non-cardiovascular specialists, are required to assist in your care, Dr. Losordo will facilitate appropriate consultation. You will continue to provide Dr. Losordo and his colleagues your current address and telephone number

16. Additional Consideration

Consent for autopsy: In the event of your death, permission for an autopsy examination will be requested in order for you doctors to gain a fuller understanding of the consequences of cell transfer. You have been told that in the event of your death, they will seek permission from your relatives to perform the autopsy and you have been asked to discuss this with your closest relatives upon entering this study.

The subject has been informed of the nature and purpose of the procedures described above including any risks involved in the research stud's performance. The subject has been asked if any questions have arisen regarding these procedures and these questions have been answered to the best of the Caritas St. Elizabeth's Medical Center's ability. A signed copy of this informed consent has been provided to the subject.

Also, any new and unforeseen information relative to the patient that may develop during the course of this research activity will be provided to the subject and the \Research/Human Subjects Committee (IRB). I will inform any referring physician9s) of any and all protocol changes, adverse events and/or safety reports.

Investigator's Signature Date

or

Representative's Signature

I have been informed about the procedures, risks, and benefits of the Research Study and agree to participate. I know that I am free to withdraw my consent and to quit the research Study at any time. My decision not to participate in this Research Study or my decision at any time to withdraw from this Research Study will not cause me any penalty or loss of benefits that I am otherwise entitled to.

I have read and understand the terms of this Consent Form and I have had the opportunity to ask questions about the Study and to discuss the Study with my doctor and other health care providers and my family and friends.

I Hereby consent to my medical records relating to this research activity to be made available to state and federal agencies (including but not limited to the Department of Health and Human Services' Food and Drug Administration (FDA), which regulates medical research activity, including this experiment. I understand that while every effort will be made to keep my identity confidential, there may be occasions when my identity must be made known to state and federal agencies at their request.

I understand that the research/Human Subjects Committee (IRB) of Caritas St. Elizabeth's Medical Center (CSEMC) has approved the solicitation of subjects to participate in this research activity.

Signature of Subject   Date   Printed Name

Name____________________________________

Or Signature of Subject's Legal Representative

Signature of Witness   Date   Printed Name

Appendix B

Heart Transplantation

The following paragraphs explain the nature of a heart transplantation program. It is hoped that the information will be of help in making a decision as to whether or not a heart transplant is appropriate for you. I owe a debt of gratitude to the University of Southern California, Keck School of Medicine for much of the information.

Here is what is involved in getting a heart transplant. A team of health professionals will do the following: (1) Find out if you are healthy enough to receive a new heart, (2) find out if you are sick enough to need a new heart, (3) prepare you for your operation (this can take a long time, and there is no guarantee that a new heart will be found), (4) perform a heart transplant operation, (5) help you stay healthy after your heart transplant.

Before you can have a heart transplant, a team of healthcare professionals will want you to think about what is likely to happen if you do get a new heart. Considerations include your physical health, your mental health, and your ability to get the transplant medications you will need.

Next, you will visit a transplant center. This is usually arranged by your doctor, but at some centers, you can make the appointment on your own. Your doctor will be asked to send your medical records to the transplant center to make sure you are taking your medicines properly and have no medical conditions that would prevent you from receiving a transplant.

At the transplant center, you will receive further testing to evaluate the following: your heart; your kidneys and gallbladder; your stomach, esophagus, and intestines; your lungs; your bladder; your teeth and gums; your prostate if you are a man; and your breasts and cervix if you are a woman. There may be other tests and examinations that the transplant team may need to perform.

Your mental health will also be evaluated. You may need to see a social worker or psychologist to make sure you do not have alcohol or drug addictions or emotional problems that could interfere with your health.

You will see an insurance specialist to evaluate your insurance to make sure you have enough coverage for the transplant and for medications after the transplant.

Once it is determined that you are a good candidate for transplantation, the transplant team will work with you to find the best transplant for you.

The transplant team will want to find out if you are sick enough to need a heart transplant. Because many heart diseases worsen slowly or respond to different therapies, the transplant center may wish to review your treatment options. They will want to determine if you have had the best possible therapy for your condition. Sometimes a new medication or therapy can make a big difference to how you feel and to the worsening of your heart disease. The transplant center does its best to make sure that you have had every chance to get better and live longer with your own heart before they will recommend a heart transplant. During the evaluation period, you will be seen as an outpatient or possibly be admitted to the transplant hospital for the evaluation. The purpose of the evaluation is to see whether transplantation would be your best option.

During the evaluation, you will meet many people and have many blood and skin tests, x-rays, and other diagnostic procedures. These tests will help the transplant team determine if transplantation would help you. Occasionally, the team discovers something that would make transplantation too dangerous for you. This could be an infection somewhere in your body, high blood pressure in your lungs, or severe thickening of your arteries.

People you will meet during your evaluation will include a transplant coordinator, cardiologist, pulmonologist, social worker, and financial counselor. In addition to these team members, you may also meet with a neurologist, psychiatrist, physical therapist, and/or other specialists. The transplant coordinator will coordinate your care and be available to answer any questions you may have. The information gathered about you from all of the tests and from each team member will help in the decision of whether transplantation is the best treatment for you.

The evaluation period is not only for the team to learn about you, but also for you and your family to meet the team and learn about transplantation and what the program has to offer you and your family. It will also give you an opportunity to learn about life with a transplanted organ and, most important, to decide if you wish to pursue transplantation. If you are accepted for transplantation, the decision to have a transplant is yours and yours alone. A commitment to do well following transplantation is necessary and a strict regimen must be adhered to. During your medical evaluation, one or more of the transplant team members will take your complete history. You will be given a physical examination that will include laboratory and diagnostic testing to determine your eligibility for transplantation. These tests may be performed at the transplant center or at a local hospital in conjunction with your primary physician.

A social worker will meet with you and your family to identify support systems, your feelings about your illness, and the possibility of transplantation. The social worker will also be able to assist you with your relocation needs. There are many resources available, both within and outside the hospital, that the social worker will be able to share with you.

You may also have an opportunity to join a group of other potential transplant recipients. The purpose of the group is to give you a chance to meet transplant candidates and recipients, address social issues that surround transplantation, and increase your knowledge about transplantation. Issues often discussed are coping with waiting for transplantation, fear of the operation itself, and insurance issues. Sometimes transplant team members will come to the group to talk about the transplant, medications, and anything else of interest to the group. There is usually a question-and-answer session afterwards. Parents and family members who come to the group find it supportive and informative. There are no fees for attending group meetings.

You may need to see the team neurologist who will examine you and discuss any past or present neurological problems that might have an effect on you after the transplantation. Neurological tests will be done as indicated by your condition and as deemed necessary by the neurologist.

A psychiatrist will conduct an exam to provide the team with an understanding of you and how you cope in your family life. The psychiatrist will assist you and your family in dealing with the stresses of a chronic disease. If you are chosen for a transplant, the psychiatrist will help with the stresses of the waiting period and transplant process.

Because exercise is vital to your well-being after a transplant, exercise will become an important part of your life after transplantation. A physical therapist will meet with you to evaluate your current physical condition and limitations, given the severity of your disease. The therapist will also be interested in the types of physical activity, if any, you enjoyed and took part in before you had your health problems. Your joint motion, muscle strength, and condition will be evaluated to make sure there are no physical problems that will affect your participation in an exercise program after transplant surgery. In addition, the physical therapist will teach energy-saving methods to help you conserve energy while waiting for a donor.

Some patients may have an examination, consultation, and/or x-rays of their teeth during the evaluation period. Patients may be cleared through their private dentist. This is done to rule out any sources of hidden or potential infection in your mouth since infections of the mouth can be very troublesome if they occur after transplantation. The dentist will make recommendations for any treatment that you may need. You may have any necessary work performed by the hospital's dentist.

Even though the waiting period can be long and you have plenty of time to anticipate surgery, most patients say they are never fully prepared when the day really comes. Most of the time, they are rushing to get to the hospital and rushing to get prepped for surgery so there is not time for last-minute preparation.

Most transplant centers offer classes on the essentials of transplantation. The classes are usually given in a lecture and open discussion format. This gives the patient and family members a chance to ask questions about anything that is not understood. All patients and interested family members are urged to attend each class in the series at least once. Patients are welcome to attend while in the hospital, or after discharge.

If you are accepted into a transplant program, a member of the team will notify you, and you will be placed on the waiting list. Your position on the list is based on your height or weight, blood type, time on the list, and the severity of your illness. The length of the list varies and may change daily.

Waiting for a suitable donor may take one week to a couple of years. Many patients and their families have described this as the most difficult part of the transplant process. Denial, fear, anxiety, and uncertainty are normal reactions experienced by patients and their families. Because this can be a frustrating and difficult time, transplant centers encourage you and your family to attend family support group sessions to help with this process.

During the waiting period, you may experience one or more hospital admissions to help control your disease. The transplant center should be notified of any change in your condition, hospitalizations, and change of address and/or telephone number. Even if you do not need to be hospitalized, you should continue to see your doctor regularly. In addition, you may need to visit the transplant center periodically so that they can monitor your condition.

Unfortunately, there are not enough hearts for every patient in need, and some people may die while waiting for a transplant. Public campaigns urging people to sign an organ donation card and let loved ones know of their wishes are very important in improving the shortage of organ donors. Donor hearts come from individuals who have been declared brain dead, usually from severe head injuries resulting from car accidents, gunshot wounds, or bleeding in the brain. The donor must be of similar height and weight as you are. The ethnic background and sex of the donor do not matter. Donor organs are given according to the severity of illness, size, blood group compatibility, and the length of time spent on the waiting list. When a heart that is a match for you becomes available, you will be contacted by phone or pager. It is important that you keep your pager on at all times and check the batteries weekly.

After you have a transplant, you may have to remain close to the hospital for a few months because of your medical needs. Although complications can occur at any time, the first three months following transplantation are the most crucial. Patients are seen weekly in the transplant clinic for the first month after discharge. Because of this need for frequent monitoring, some patients who live more than sixty-five miles from the transplant center may have to take up temporary residence near the transplant center.

When a suitable donor is found, one of the members of the transplant team will notify you by telephone. It is important that you always leave a number where you can be reached if you are away from home. As a rule of thumb, the transplant center will usually call your home first. Purchase of a pager is recommended as a means of contacting you. Otherwise, the transplant team will have to rely on your work, relative, or cell phone number as a means of contacting you when a heart becomes available.

Just before surgery, you will have blood drawn so that the team has up-to-date information about the key components of your blood. You will also be asked to give a urine specimen. In order to reduce the possibility of infection, you will have your chest thoroughly washed with a special cleansing solution, and your chest may be shaved. Finally, you will receive medications that will help you relax and begin to make you feel sleepy.

When a heart that meets your requirements is located, you will be called into the hospital by the nurse coordinator. The transplant doctors will be checking the donor heart while you are being evaluated and started on medications in preparation for transplantation. If the donor organ is good, you will then be taken to the operating room and put to sleep with an anesthetic. One of the transplant surgeons will begin the process of preparing your chest cavity for removal of your heart.

The surgeon will begin by exposing the chest cavity through a cut in the ribcage. The surgeon will then open the pericardium (a membrane that covers the entire heart) in order to remove your diseased heart. The back part of your left atrium will be left in place, but the rest of the heart will be removed. Your new heart will be carefully trimmed and sewn to the remaining parts of your old heart. This transplant method is called an "orthotopic" procedure. This is the most common method used to transplant hearts.

You will be given medications both before and during the operation to prevent rejection of the new heart. After the operation, you will be taken to a special unit of the hospital for recovery. You will stay in the hospital until your doctor believes you are ready to go home. How long you stay in the hospital will depend on the following factors: how well your new heart is working and your ability to take care of your new heart transplant.

You will wake up in the intensive care unit (ICU) after surgery. Around you, there will be many machines, tubes, and people. You will hear bells, beeps, and talking. You will have a tube in your mouth and throat that helps you to breathe but keeps you from being able to talk. You cannot eat or drink anything while the tube is in your mouth. You will have intravenous lines in the veins of your arms and your neck that will give you fluids and medications. You will feel a tube (one or two) coming from your chest which is draining fluid that can collect there. A large dressing will be covering the wound on your chest from your surgery. You will notice wires on your chest which will connect to a monitor so your nurse can see how your heart is working. A catheter will continually empty your bladder.

Once you wake up and are able to breathe adequately, the ventilator tubes will be removed. You will have some pain after the surgery and will receive pain medications either intravenously or by mouth. As you recover, you will be transferred to a step-down unit and then to a regular room.

While in the hospital, frequent blood tests, including those measuring drug levels, electrolytes, and kidney function, are performed. The transplant team will be watching for possible post-transplant complications, including infection and rejection of the heart. You will receive several heart biopsies after the transplant to monitor for rejection. Heart biopsies involve inserting a tube holder into a vein in the neck or groin. Through this tube, a bioptome (biopsy device) is inserted into the heart, and small samples are taken. A pathologist reviews these to see if rejection is present. The frequency of biopsies depends on the time elapsed from transplant, your rejection history, and the protocols followed at each transplant center.

While in the hospital, you will receive counseling from a dietitian, a physiotherapist, and a pharmacist to prepare you for your return home. It is important to have someone with you to help remember all of the information. Many medications are required after the transplant, each with a specific purpose. There are medications to lower rejection (immuno-suppressants), medications to treat cholesterol (statins), medications to prevent infection, and sometimes medications to treat high blood pressure and other complications that may occur after transplant. These medications are often adjusted depending on side effects and the presence or absence of rejection. All medications have potential side effects, some more serious than others, although most people tolerate the medications well. Minor side effects may include rash and stomach upset. Serious side effects may include an increased risk of infections, diabetes, osteoporosis (thinning of the bones), high blood pressure, kidney disease, and the development of cancer. Lymphoma and skin cancer can occur after the transplant and are related to anti-rejection medication and its effects on the immune system.

Appendix C

End Notes

1. Stipp, D. "The first blockbuster treatment using stem cells may emerge sooner than most experts dreamed." *Fortune*. Nov 29, 2004.
2. Harasim, P. "Stem cells give heart patient new lease on life." reviewjournal.com. Feb 14, 2005.
3. Caffrey, S.L., P.J. Willoughby, and L.A. Becker. *New England Journal of Medicine*. 2002. 347:1242-7.
4. "Transmyocardial laser revascularization." Texas Heart Institute at St. Luke's Episcopal Hospital.
5. Fogoros, R.N. "Secret Cardiology—EECP: Your guide to heart disease/cardiology." Undated.
6. "TheraVitae to present stem cell interim trial results at American Association Conf." Press release. Oct 13, 2005.
7. "Stem cells and stem cell transplantation." National Library of Medicine.
8. "Stem cells: scientific progress and future research." National Institutes of Health.

9. “Stem cell transplant.” Mayo Foundation for Medical Education and Research.
10. “Stem cell information.” National Institutes of Health.
11. Muller, A.C., et al. “Long term survival and transplantation of haemopoietic stem cells for immunodeficienceies.” *Lancet.* Feb 15, 2003.
12. Wollert, K.C., et al. “Clinical applications of stem cells for the heart.” Department of Cardiology and Angiology, Hanover Medical School, Hanover, Germany.
13. Ho, M.W. “Adult stem cells proving themselves in the clinic.” Institute of Science in Society. Jan 1, 2005.
14. Willerson, J.T. “Stem cell research in Texas and Brazil.” University of Texas Health Science Center. May 13, 2005.
15. “Brazil begins major stem cell study to treat heart disease.” stemcellnews.com. Jun 10, 2005.
16. “Stem cell therapy sparks hope in ailing hearts.” Reuters. Jan 8, 2006.
17. Rosengart, T.K. “Gene therapy for cardiovascular disease.” Evaston Northwestern Healthcare. Undated.
18. “Adult stem cells from adipose tissue could save lives.” *Stem Cell Research News.* Sep 5, 2005.
19. “Neurosurgeons looking at stem cells from skin to fight brain tumors.” Congress of Neurological Surgeons. Oct 10, 2004.
20. Gargett, C. “Discovery of adult stem cells in uterus can be grown into bone, muscle etc.” Monash Institute of Medical Research. Jul 19, 2005.
21. Fumento, M. “Adult stem cells provide new life for livers.” *The New Scientist.* Oct 20, 2005.
22. “Ultimate stem cell discovered.” *New Scientist*, print edition. Jan 23, 2002.

23. Johnston, L. "Russian stem cell therapy healing therapies." info/russianstemcell.htm. Undated.

24. Yann, B., et al. "Application of stem cell research to epithelial repair." *Advances in Stem Cell Research.* Lausanne, Switzerland. Sep 10, 2006.

25. Wilson, B.E. "5-Alpha-Reductase deficiency." emedicine.com. Undated.

26. Baier, A.D., et al. "Stem cells reduce stroke size and damage." Research.online, research.usf.edu. Jul 28, 2006.

27. "Former county judge Mills Lane to get stem cell injections in Ukraine." *Reno-Gazette.* Jun 17, 2004.

28. Patschan, et al. "Dynamics of mobilization and homing of endothelial progenitor cells after acute..." *American Journal of Physiology.* 2006.

29. Oswald, J., et al. "Mesenchymal stem cells can be differentiated into endothelial cells in vitro." stemcells.com. Undated.

30. *FDA Consumer Magazine.* Sep - Oct 2003 issue.

31. "States that require health plans to cover patient care costs in clinical trials." National Cancer Institute. Dec 19, 2002.

Appendix D

Glossary

Adipose tissue – Fat.

Adult stem cells – An undifferentiated cell (unchanged) which can multiply and change into almost any other type of cell.

Aerobic exercise – Exercise that increases the need for oxygen.

Angina – Chest pain from heart.

Angiogenesis – The development of blood vessels.

Angioplasty – Surgical repair of a blood vessel by insertion of a balloon-tipped catheter.

Apheresis – A procedure in which blood is drawn from a donor and separated into its parts.

Balloon angioplasty – See Angioplasty.

Blinded test – A test in which the subject does not know if a medication is received or a placebo.

Bone marrow – The soft, fatty vascular material that fills most bones.

Browser – A program that accesses and displays files and other data available on the Internet.

Bypass surgery – A surgical procedure in which a graft is attached to an artery above and below the site of obstruction to provide normal blood flow to the diseased vessel.

Cardiac arrest – Cessation of heart function.

Cardiac intensive care – Continuous and closely monitored healthcare that is provided to critical heart patients.

Cardiac rehabilitation – A formal program consisting of dietary regimen, medication, and progressive exercise, directed toward improving cardiac function.

CAT scan – A sectionalized view of the body constructed by computer tomography, also called a CT scan.

Catheter – A long, extremely narrow tube used in cardiac diagnostic and therapeutic procedures to introduce fluids or devices into blood vessels.

CD34-positive stem cells – Hematopoietic stem cell, endothelial progenitor.

Cirrhosis of the liver – A chronic disease of the liver caused by replacement of normal tissue with fibrous tissue.

Clinical trial – A scientific test of the effectiveness and safety of a therapeutic agent using human subjects.

Collateral circulation – The development of connections around an obstruction to blood flow by the growth of small arteries above and below the obstruction.

Congestive heart failure – Heart failure caused by venous congestion producing prominent neck veins, fluid retention in the lungs, liver enlargement, and edema (water) in the legs.

Cord blood stem cells – Stem cells found in blood in the umbilical cord at the time of delivery.

Coronary artery – Main arteries and their associated branches that supply the heart with the oxygen and nutrients it needs to function.

Defibrillator – An electrical device used to counteract fibrillation of the heart muscles and restore normal beat by applying a shock.

Echocardiogram – An image of the heart produced by ultrasonic imaging.

Ejection fraction – That part, or fraction, of all blood in the ventricle (main pumping chamber) that is actually ejected at each heartbeat.

EKG – A recording of the electrical impulses of the heart, usually in the form of an electrocardiogram.

Embryonic stem cells – Primitive cells from embryos that have the potential to become a wide variety of specialized cells.

Endothelial tissue – The innermost layer of tissue lining a blood vessel.

FDA – U.S. Food and Drug Administration.

Fetal stem cells – Stem cells found in the blood of fetuses.

Gene therapy – The treatment of certain disorders, especially those caused by genetic anomalies or deficiencies, by introducing specially engineered genes into a patient's cells.

Heart catheterization – Introduction of a catheter, usually through an artery in the groin, to examine or treat the heart.

Heart pump – A pump utilized to increase pumping capacity of a diseased heart.

Heart-lung machine – A mechanical pump used during heart surgery to shunt blood away from the heart, oxygenate it, and return it to the body, thereby maintaining blood circulation.

Hemopoietic stem cells – A stem cell from which all red and white blood cells evolve.

Hyperbaric chamber – A compartment capable of high-pressure oxygenation used to treat anaerobic infections.

Informed consent – Consent by a patient to a surgical or medical procedure or participation in a clinical study after achieving an understanding of the relevant medical facts and risks involved.

Institutional review board – Reviews the protocol of a clinical study to make sure the clinical trial is conducted fairly and participants are not likely to be harmed.

Internet – An interconnected system of networks that connects computers around the world via the TCP/IP protocol.

Ischemia – Tissue anemia (as of heart muscle) resulting from lack of flow of arterial blood.

IV – An apparatus for providing intravenous injections.

Left ventricle – The chamber on the left side of the heart that receives the blood from the left atrium and contracts to force it into the aorta.

Macular degeneration – A gradual loss of central vision characterized by spots of pigmentation and causing a reduction or loss of central vision.

Mesenchymal stem cells – Cells from the embryonic and adult connective tissue. A number of cell types come from mesenchymal stem cells, including chondrocytes, which produce cartilage.

MRI – Magnetic resonance imaging which produces images of slices of the body.

Multiple adult progenitor cells – Adult stem cells.

Nephrologist – A specialist who is expert in the treatment of kidney disease.

Neupogen – A trade name for a form of drug to move cells from the bone marrow to the bloodstream.

Nitroglycerin – A pill or liquid medication that dilates coronary arteries to reduce angina (chest pain).

Nuclear stress test – A test in which a small amount of radioactive substance is injected into the bloodstream, and a medication is given to have a similar effect on the heart as walking on a treadmill. A scan of the heart is made before and after the medication is given. The test will indicate areas of damage or reduced blood flow in the heart.

Pacemaker – An electronic device that takes over the pacing or rhythm of the heart.

Placebo – A substance containing no medication and prescribed to a patient to reinforce the expectation of getting well.

Plaque – A raised patch or swelling on the inner surface of an artery produced by fatty deposits.

Progressive care units – Areas of a hospital for patients who need cardiac monitoring but do not require intensive care and monitoring.

Protocol – A plan for a course of medical treatment or for a scientific experiment.

Randomized – To make random in arrangement, especially in order to control the variable in a scientific experiment.

Refractory angina – Occurrence of frequent angina attacks uncontrolled by optimal drug therapy.

Respirator tube – A device that supplies oxygen or a mixture of oxygen and carbon dioxide for breathing.

Retinitis pigmentosa – A hereditary degenerative disease of the retina characterized by night blindness, pigmentary changes within the retina, and eventual loss of vision.

Rheumatoid arthritis – A chronic disease marked by stiffness and inflammation of the joints, weakness, loss of mobility, and deformity.

Search engine – A software program that searches a database and gathers and reports information available on the Internet.

Staphylococcus – A spherical parasitic bacterium usually occurring in grape-like clusters and causing boils, septicemia, and other infections.

Stem cells – Cells that have the ability to divide for indefinite periods and to change into specialized cells.

Stent – A short, narrow metal or plastic tube often in the form of a mesh that is inserted into an artery to keep a previous blockage open.

Stromal stem cells – Cells from bone marrow that are adult stem cells that contribute to the regeneration of tissues including bone, cartilage, fat, and muscle.

TMLR – A type of surgery that uses a laser to make tiny channels through the heart muscle to improve blood circulation to the heart muscle.

Treadmill test – A patient walks on a treadmill while blood pressure and EKG are monitored. The treadmill test evaluates the heart's response to exercise.

UNOS – United Network for Organ Sharing, a non-profit organization that coordinates U.S. organ transplant activities.

M
Maine. Office
Alcohol and drug

Printed in the United States
76246LV00001B/175-1749

9 781600 080371